Wild Strawberry
Resources in China

中国野生草莓资源

雷家军
薛　莉　主编

北方联合出版传媒（集团）股份有限公司
辽宁科学技术出版社

图书在版编目（CIP）数据

中国野生草莓资源 / 雷家军, 薛莉主编. -- 沈阳 : 辽宁科学技术出版社, 2025.2. -- ISBN 978-7-5591-4061-6

Ⅰ. S668.402.4

中国国家版本馆CIP数据核字第2025G4J480号

出版发行：辽宁科学技术出版社
　　　　　（地址：沈阳市和平区十一纬路25号　邮编：110003）
印 刷 者：辽宁鼎籍数码科技有限公司
经 销 者：各地新华书店
幅面尺寸：210mm×285mm
印　　张：12
字　　数：300千字
出版时间：2025 年 2 月第 1 版
印刷时间：2025 年 2 月第 1 次印刷
责任编辑：陈广鹏
封面设计：周　洁
责任校对：栗　勇　李　婵

书　　号：ISBN 978-7-5591-4061-6
定　　价：158.00 元

联系热线：024-23280036
邮购热线：024-23284502
http://www.lnkj.com.cn

主　编

雷家军　薛　莉

副主编

余　红　周厚成　韩永超　杨瑞华

参　编

张运涛　赵密珍　乔玉山　秦国新　张体操　乔　琴
李洪雯　冯嘉玥　王桂霞　孙　健　王　冲　屈连伟
毕晓颖　代汉萍　赵　珺　郑　洋　毕蒙蒙　曲　波
马迎杰　岳静宇　贾宇美凤　蒋　姝　杜秋玲　马睿一
纪　艺　屈　麟　张　伟　罗刚军　郭瑞雪　赵　艳
武　建　陈　红　邓小敏　庄子涵　徐铭毅　王志刚
王洪安　张亚艳　旦真次仁　付本涛　栾绍武　白　鹏
刘丰亮　王丹菲　韩　玲　梁　潇　马福刚　肖文斐
阮继伟　姜兆彤　马廷东　杨远杰　赵文平　许　珂

雷家军 教授
沈阳农业大学园艺学院

Prof. Jiajun Lei
College of Horticulture, Shenyang Agricultural University

雷家军，1966年10月生，沈阳农业大学园艺学院教授，留日博士后，博士研究生导师。中国园艺学会草莓分会副理事长、辽宁省特聘教授。从事草莓种质资源、遗传育种、生产栽培、生物技术等方面研究36年。主持国家自然科学基金3项及其他国家、省（部）市级课题50项。主持培育草莓等园艺植物新品种16个。发表论文302篇，其中SCI、CPCI收录60篇；出版著作34部，其中主编《中国果树志·草莓卷》《中国草莓》等重要著作。

Dr. Jiajun Lei, born in October 1966, professor at College of Horticulture, Shenyang Agricultural University, postdoctoral fellow in Japan, and doctoral supervisor. Vice Chairman of Strawberry Section of Chinese Society for Horticultural Science, and Special Professor of Liaoning Province. Engaged in researches on strawberry germplasm resources, genetic breeding, cultivation and biotechnology for 36 years. Took charge of 3 National Natural Science Foundation Projects and other 50 scientific research projects. Bred 16 new cultivars of horticultural plants including strawberry. Published 302 articles, including 60 indexed by SCI or CPCI. Published 34 books, including some important works such as *China Fruit Tree Records, Volume Strawberry* and *Strawberries in China*.

薛 莉 副教授
沈阳农业大学园艺学院

Associate Prof. Li Xue
College of Horticulture, Shenyang Agricultural University

薛莉，1988年10月生，博士，副教授，硕士研究生导师。沈阳农业大学园艺学院教师，沈阳农业大学天柱山青年骨干教师人才，沈阳市拔尖人才，中国园艺学会草莓分会副秘书长。主要从事草莓等园艺植物种质资源、遗传育种、生物技术等方面的研究。主持国家自然科学基金1项及主持参加其他国家、省（部）市级科研课题20余项。主持或参加培育草莓等园艺植物新品种15个。发表论文52篇，其中SCI、CPCI收录35篇，主编《有机草莓栽培实用技术》，副主编全国高等农林院校教材《浆果栽培学》（第二章：草莓）。

Dr. Li Xue, born in October 1988, associate professor, master tutor. Teacher at College of Horticulture, Shenyang Agricultural University, Young Core Teacher of Tianzhu Mountain in Shenyang Agricultural University. Shenyang top-notch talent. Vice secretary gereral of Strawberry Section of Chinese Society for Horticultural Science. Engaged in researches on germplasm resources, genetic breeding and biotechnology of horticultural plants including strawberry. Took charge of 1 National Natural Science Foundation Project and other more than 20 scientific research projects. Bred 15 new cultivars of horticultural plants including strawberry. Published 52 articles, including 35 indexed by SCI or CPCI. A chief editor of *Cultivation Technology of Organic Strawberry*, a vice-chief editor of the national agriculture and forestry colleges textbook *Science of Berry Cultivation* (*Chapter II: Strawberry*).

前　言

　　草莓浆果芳香多汁，酸甜适口，营养丰富，素有"水果皇后"的美称，是最受人们喜爱的水果之一。据联合国粮农组织（FAO）统计，自1994年以来，中国草莓年产量一直居世界第一位，2023年中国草莓栽培面积和产量分别为15.63万hm²（234.45万亩）和421.67万t，年产量约占全世界的1/3。

　　中国是世界上草莓野生种类资源最丰富的国家，全世界草莓属26个种中（25个为野生种、1个为栽培种），我国自然分布15个野生种，占一半以上。这15个种包括10个二倍体种和5个四倍体种。10个二倍体种是五叶草莓、黄毛草莓、森林草莓、西藏草莓、裂萼草莓、绿色草莓、东北草莓、中国草莓、台湾草莓、峨眉草莓，5个四倍体种是东方草莓、西南草莓、伞房草莓、纤细草莓、高原草莓。而且我国分布有一些特有的种类，例如五叶草莓、中国草莓、峨眉草莓、台湾草莓等仅分布在我国，全世界5个四倍体种也几乎全部分布在我国。此外，近年来我们还发现我国东北分布有自然五倍体野生草莓。我国这些珍贵野生资源不仅可以直接加以利用，而且是拓宽栽培草莓遗传基础的重要基因资源，有重要的科研价值和利用价值。

　　我们用中英文双语编写了《中国野生草莓资源》一书，对我国分布的野生草莓资源种类和利用进行了全面系统的介绍。全书共分八章，包括中国野生草莓资源的考察与收集、保存、分布、分类、特点、生境、季相、利用。本书以彩色图片为主，内容简洁，图文并茂，排版精美，条理清晰，学术性和科普性强，是一本研究中国草莓属植物的工具书，可供草莓科研人员、技术员、种植户、院校学生、爱好者等参考利用。

　　本书的出版得到了全国草莓同行的大力帮助，在此深表谢意！由于时间和水平有限，本书一定存在很多不足之处，希望大家不吝赐教，以便我们及时更正。

　　本书得到了国家自然科学基金项目的资助（项目编号：32372652），特此鸣谢！

雷家军　薛莉

2025年2月1日

Introduction

Strawberry is known as 'the Queen of Berries' for its strong aroma, delicious favor and rich nutrition. According to the statistics of the Food and Agriculture Organization of the United Nations (FAO), China's annual strawberry output has ranked first in the world since 1994. In 2023, China's strawberry cultivation area and output was 156,300 hm^2 and 4,216,700 tons respectively, accounting for about one-third of the world's annual output.

China has more wild strawberry resources than any other country in the world. Of about 26 recognized *Fragaria* species (including 25 wild species and 1 cultivated species), 15 wild species are distributed in China, including 10 diploid species: *F. pentaphylla* Lozinsk., *F. nilgerrensis* Schlechtendal ex J. Gay, *F. vesca* L., *F. nubicola* Lindl., *F. daltoniana* Gay, *F. viridis* Duch., *F. chinensis* Lozinsk., *F. mandschurica* Staudt, *F. hayatai* Staudt, and *F. emeiensis* Jia J. Lei, and 5 tetraploid species: *F. orientalis* Lozinsk, *F. moupinensis* (Franch) Card., *F. corymbosa* Lozinsk., *F. gracilis* A. Los. and *F. tibetica* Staudt et Dickoré. There are some unique species only distributed in China, such as *F. pentaphylla*, *F. chinensis*, *F. emeiensis*, and *F. hayatai*, and all the 5 tetraploid species in the world are almost only distributed in China. Besides, the wild pentaploid strawberry genotype has been discovered in Northeast China in recent years. These precious wild resources native to China can not only be used directly, but also are important resources to broaden the genetic basis of modern cultivated strawberry.

We have written the book *Wild Strawberry Resources in China* in both Chinese and English to give a systematic introduction of the wild *Fragaria* species native to China and their utilization. This book is a guide of the genus *Fragaria* distributed in China. It consists of 8 chapters, including the investigation and collection, conservation, distribution, taxonomy, characteristics, habitat, season aspect, and protection and utilization of wild strawberry resources native to China. We hope this book will be vivid and easy to read as it is richly illustrated with a clear layout. It can be used as a valuable reference book by strawberry researchers, technicians, growers, college students and enthusiasts for its highly academic value and popularization value.

The publication of this book has been strongly and kindly supported by Chinese strawberry colleagues. Due to the limited time and skills there might be a lot of mistakes in this book. Please do not hesitate to contact us so that we can revise them in time.

This work was supported by the National Natural Science Foundation of China (No. 32372652).

Jiajun Lei and Li Xue

February 1, 2025

目 录

Contents

WILD STRAWBERRY RESOURCES IN CHINA

中国野生草莓资源的考察与收集
Investigation and collection of wild strawberry resources in China

　　我国对国内野生草莓资源考察收集工作进行得较晚，一些科研单位从20世纪80年代开始对野生草莓资源进行考察收集工作，如沈阳农业大学、北京市农林科学院、江苏省农业科学院等单位，其他一些科研单位如中国农业科学院郑州果树研究所、湖北省农业科学院经济作物研究所、四川省农业科学院园艺所、云南农业大学、杭州市农业科学院、黑龙江省农业科学院园艺分院、西北农林科技大学、贵州省农业科学院园艺所、甘肃省农业科学院林果花卉所、昆明市农业科学院、蛟河草莓研究所、重庆元邦农业发展有限公司等也进行了较多考察收集，并保存了大量野生草莓资源。

　　中国园艺学会草莓分会在张运涛理事长、雷家军副理事长、赵密珍副理事长的倡导下，于2016年开始组织草莓专业考察队，对我国境内野生草莓分布区开展了较大规模的深入考察，先后考察了四川、西藏、云南、贵州、甘肃、黑龙江、湖北、吉林、陕西、山西等省区，对厘清我国野生草莓资源状况做出了很大贡献。

The investigation and collection of wild strawberry resources in China was carried out late, and several scientific research institutes have begun to carry out until the 1980s, such as Shenyang Agricultural University, Beijing Academy of Agriculture and Forestry Sciences, and Jiangsu Academy of Agricultural Sciences. Other research institutions have also investigated, collected and preserved some wild strawberry resources, such as Zhengzhou Fruit Research Institute of Chinese Academy of Agricultural Sciences, Industrial Crops Institute of Hubei Academy of Agricultural Sciences, Horticulture Institute of Sichuan Academy of Agricultural Sciences, Yunnan Agricultural University, Hangzhou Academy of Agricultural Sciences, Horticulture Institute of Heilongjiang Academy of Agricultural Sciences, Northwest A&F University, Horticulture Institute of Guizhou Academy of Agricultural Sciences, Fruit and Floriculture Institute of Gansu Academy of Agricultural Sciences, Kunming Academy of Agricultural Sciences, Jiaohe Strawberry Research Institute, and Chongqing Yuanbang Agricultural Development Co., Ltd.

Under the initiative of president Yuntao Zhang, vice-president Jiajun Lei and vice-president Mizhen Zhao, Strawberry Section of Chinese Society for Horticultural Science has begun to organize the several professional wild strawberry expeditions since 2016 and carried out some large-scale and in-depth investigation of wild strawberry resources throughout China, including Sichuan, Xizang, Yunnan, Guizhou, Gansu, Heilongjiang, Hubei, Jilin, Shaanxi, Shanxi and other provinces. It has made a major contribution to clarifying the status of wild strawberry resources in China.

1.1 沈阳农业大学

Shenyang Agricultural University

沈阳农业大学是我国最早开展野生草莓资源研究和利用的单位，从1980年开始进行考察、收集、分类、鉴定、评价、利用工作。在我国著名草莓专家邓明琴教授收集野生草莓的基础上，雷家军教授从1991年开始，持续不断地开展了对国内外野生草莓资源的收集保存工作，尤其是对中国野生草莓资源的考察收集，先后实地考察了辽宁、吉林、黑龙江、新疆、山西、青海、内蒙古、四川、西藏、云南、贵州、甘肃、湖北、陕西等省区，重点对我国的天山、长白山、秦岭、大兴安岭、青藏高原、云贵高原、神农架等地进行了考察，发现了一些新种、变种和类型，基本厘清了我国自然分布的草莓属植物种类和分布情况。

Shenyang Agricultural University is the first institute to carry out the research and utilization of wild strawberry resources in China, and began their investigation, collection, classification, identification, evaluation and utilization in 1980. Based on the wild strawberry collection by late Prof. Mingqin Deng, the most famous strawberry expert in China, Prof. Jiajun Lei has continuously carried out the collection of wild strawberry resources at home and abroad, especially the Chinese strawberry resources since 1991. He has visited Liaoning, Jilin, Heilongjiang, Xinjiang, Shanxi, Qinghai, Inner Mongolia, Sichuan, Xizang, Yunnan, Guizhou, Gansu, Hubei, Shaanxi and other provinces, focusing on Mount Tianshan, Mount Changbai, Mount Qinling, Mount Greater Hinggan, Qinghai-Xizang Plateau, Yunnan-Guizhou Plateau, Shennongjia Forest Region and other sites. Some new species, varieties and types were discovered, which further clarified the species and distribution of wild strawberry in China.

1988年邓明琴教授（右）和洪建源教授（左）夫妇在新疆天山考察野生草莓
Prof. Mingqin Deng (right) and her husband Prof. Jianyuan Hong (left) investigated the wild strawberry resources in Mount Tianshan, Xinjiang Autonomous Region in 1988.

2008年7月雷家军教授在山西五台山考察野生草莓
Prof. Jiajun Lei investigated the wild strawberry resources in Mount Wutai, Shanxi Province in July, 2008.

2010年7月雷家军教授在青海西宁考察野生草莓
Prof. Jiajun Lei investigated the wild strawberry resources in Xining City, Qinghai Province in July, 2010.

2010年8月雷家军教授在黑龙江牡丹江考察野生草莓
Prof. Jiajun Lei investigated the wild strawberry resources in Mudanjiang City, Heilongjiang Province in August, 2010.

2015年5月雷家军教授在新疆塔城考察野生草莓
Prof. Jiajun Lei investigated the wild strawberry resources in Tacheng City, Xinjiang in May, 2015.

2016年5月雷家军教授与李洪雯博士（左）在四川成都考察野生草莓
Prof. Jiajun Lei and Dr. Hongwen Li (left) investigated the wild strawberry resources in Chengdu City, Sichuan Province in May, 2016.

2016年6月雷家军教授与旦真次仁（左）在西藏林芝考察野生草莓
Prof. Jiajun Lei and Mr. Ciren Danzhen (left) investigated the wild strawberry resources in Nyingchi Prefecture, Xizang in June, 2016.

2017年5月雷家军教授在辽宁丹东考察野生草莓
Prof. Jiajun Lei investigated the wild strawberry resources in Dandong City, Liaoning Province in May, 2017.

2017年6月雷家军教授在云南迪庆考察野生草莓
Prof. Jiajun Lei investigated the wild strawberry resources in Diqing Prefecture, Yunnan Province in June, 2017.

2017年7月雷家军教授与张运涛研究员（前）在河北涞源考察野生草莓

Prof. Jiajun Lei and Prof. Yuntao Zhang (front) investigated the wild strawberry resources in Laiyuan County, Hebei Province in July, 2017.

2018年6月雷家军教授与赵密珍研究员（右）在甘肃岷县考察野生草莓

Prof. Jiajun Lei and Prof. Mizhen Zhao (right) investigated the wild strawberry resources in Minxian County, Gansu Province in June, 2018.

2018年5月雷家军教授在贵州梵净山考察野生草莓
Prof. Jiajun Lei investigated the wild strawberry resources in Mount Fanjing, Guizhou Province in May, 2018.

2018年6月雷家军教授在甘肃天水考察野生草莓
Prof. Jiajun Lei investigated the wild strawberry resources in Tianshui City, Gansu Province in June, 2018.

2019年6月雷家军教授、杨瑞华副研究员（中）、张伟博士（左）在黑龙江林口考察野生草莓
Prof. Jiajun Lei, Associate Prof. Ruihua Yang (middle) and Dr. Wei Zhang (left) investigated the wild strawberry resources in Linkou County, Heilongjiang Province in June, 2019.

2021年6月薛莉博士在湖北神农架考察野生草莓
Dr. Li Xue investigated the wild strawberry resources in Shennongjia Forest Region, Hubei Province in June, 2021.

2023年5月雷家军教授与薛莉博士（右）、贾宇美凤硕士（左）在陕西略阳考察野生草莓
Prof. Jiajun Lei, Dr. Li Xue (right) and Yumeifeng Jia (left) investigated the wild strawberry resources in Lueyang County, Shaanxi Province in May, 2023.

2023年6月雷家军教授与薛莉博士（左）、岳静宇博士（右）在湖北神农架考察野生草莓
Prof. Jiajun Lei, Dr. Li Xue (left) and Dr. Jingyu Yue (right) investigated the wild strawberry resources in Shennongjia Forest Region, Hubei Province in June, 2023.

2024年7月雷家军教授与杨瑞华副研究员（左）、薛莉博士（右）在黑龙江尚志考察野生草莓
Prof. Jiajun Lei, Associate Prof. Ruihua Yang (left) and Dr. Li Xue (right) investigated the wild strawberry resources in Shangzhi City, Heilongjiang Province in July, 2024.

2024年7月薛莉博士（右）和张伟博士（左）在山西沁源考察野生草莓
Dr. Li Xue (right) and Dr. Wei Zhang (left) investigated the wild strawberry resources in Qinyuan County, Shanxi Province in July, 2024.

1.2 北京市农林科学院林业果树研究所

Forestry and Fruit Research Institute, Beijing Academy of Agriculture and Forestry Sciences

2016年5月张运涛研究员在四川成都考察野生草莓
Prof. Yuntao Zhang investigated the wild strawberry resources in Chengdu City, Sichuan Province in May, 2016.

2023年6月孙健博士在湖北神农架考察野生草莓
Dr. Jian Sun investigated the wild strawberry resources in Shennongjia Forest Region, Hubei Province in June, 2023.

2019年6月张运涛研究员团队在黑龙江尚志考察野生草莓（从左至右：高用顺、李睿、常琳琳、张运涛、孙瑞、王桂霞）
Prof. Yuntao Zhang's team investigated the wild strawberry resources in Shangzhi City, Heilongjiang Province in June, 2019. (From left to right: Yongshun Gao, Rui Li, Linlin Chang, Yuntao Zhang, Rui Sun, and Guixia Wang)

2017年6月张运涛研究员团队在云南迪庆考察野生草莓（从左至右：常琳琳、孙瑞、董静、王桂霞）
Prof. Yuntao Zhang's team investigated the wild strawberry resources in Diqing Prefecture, Yunnan Province in June, 2017. (From left to right: Linlin Chang, Rui Sun, Jing Dong, and Guixia Wang)

1.3 江苏省农业科学院果树研究所
Fruit Research Institute, Jiangsu Academy of Agricultural Sciences

2018年6月赵密珍研究员在甘肃岷县考察野生草莓
Prof. Mizhen Zhao investigated the wild strawberry resources in Minxian County, Gansu Province in June, 2018.

2010年5月乔玉山教授在四川甘孜考察野生草莓
Prof. Yushan Qiao investigated the wild strawberry resources in Ganzi Prefecture, Sichuan Province in May, 2010.

2023年6月蔡伟健博士在湖北神农架考察野生草莓
Dr. Weijian Cai investigated the wild strawberry resources in Shennongjia Forest Region, Hubei Province in June, 2023.

2023年5月赵密珍研究员团队在陕西略阳考察野生草莓（从左至右：关玲、赵密珍、蔡伟健、庞夫花）
Prof. Mizhen Zhao's team investigated the wild strawberry resources in Lueyang County, Shaanxi Province in May, 2023. (From left to right: Ling Guan, Mizhen Zhao, Weijian Cai, and Fuhua Pang)

1.4 中国农业科学院郑州果树研究所

Zhengzhou Fruit Research Institute, China Academy of Agricultural Sciences

周厚成研究员2016年5月在四川成都（左）和2017年6月在云南迪庆（右）考察野生草莓
Prof. Houcheng Zhou investigated the wild strawberry resources in Chengdu City, Sichuan Province in May, 2016 (left) and in Diqing Prefecture, Yunnan Province in June, 2017 (right).

1.5 湖北省农业科学院经济作物所

Industrial Crops Institute, Hubei Academy of Agricultural Sciences

2023年5月韩永超研究员在甘肃康县考察野生草莓
Prof. Yongchao Han investigated the wild strawberry resources in Kangxian County, Gansu Province in May, 2023.

1.6 四川省农业科学院园艺研究所
Horticulture Research Institute, Sichuan Academy of Agricultural Sciences

2023年7月李洪雯研究员在新疆特克斯考察野生草莓
Prof. Hongwen Li investigated the wild strawberry resources in Tekes County, Xinjiang Autonomous Region in July, 2023.

1.7 云南农业大学
Yunnan Agricultural University

2022年6月乔琴教授带领研究生在云南白马雪山考察野生草莓（从左至右：曹强、兰根倩、乔琴、吴明钊）
Prof. Qin Qiao and her graduate students investigated the wild strawberry resources in Baima Snow Mountain, Yunnan Province in June, 2022. (From left to right: Qiang Cao, Genqian Lan, Qin Qiao, and Mingzhao Wu)

1.8 杭州市农业科学院生物技术研究所
Institute of Biotechnology, Hangzhou Academy of Agricultural Sciences

2024年7月余红研究员在山西沁源考察野生草莓
Prof. Hong Yu investigated the wild strawberry resources in Qinyuan County, Shanxi Province in July, 2024.

1.9 黑龙江省农业科学院园艺分院
Horticultural Branch, Heilongjiang Academy of Agricultural Sciences

2019年6月杨瑞华副研究员在黑龙江林口考察野生草莓
Associate Prof. Ruihua Yang investigated the wild strawberry resources in Linkou County, Heilongjiang Province in June, 2019.

1.10 西北农林科技大学
Northwest A&F University

2021年5月冯嘉玥副教授在陕西略阳调查收集资源
Associate Prof. Jiayue Feng investigated the wild strawberry resources in Lueyang County, Shaanxi Province in May, 2021.

1.11 贵州省农业科学院园艺研究所
Horticultural Institute, Guizhou Academy of Agricultural Sciences

贵州省农业科学院园艺所草莓团队2018年5月在贵州六盘水考察野生草莓（左2、3、4分别为杨仕品、钟霈霖、乔荣）
The strawberry research group from Institute of Horticulture, Guizhou Academy of Agricultural Sciences investigated the wild strawberry resources in Liupanshui City, Guizhou Province in May, 2018. (No. 2, 3 and 4 on the left are Shipin Yang, Peilin Zhong, and Rong Qiao respectively.)

1.12 甘肃省农业科学院林果花卉研究所
Fruit and Floriculture Research Institute, Gansu Academy of Agricultural Sciences

2020年6月王卫成副研究员在甘肃兰州考察野生草莓
Associate Prof. Weicheng Wang investigated the wild strawberry resources in Lanzhou City, Gansu Province in June, 2020.

1.13 昆明市农业科学院
Kunming Academy of Agricultural Sciences

2017年7月陈杉艳高级农艺师（右2）在云南昆明考察草莓野生资源
Senior Agronomist Shanyan Chen (No. 2 on the right) investigated the wild strawberry resources in Kunming City, Yunnan Province in July, 2017.

1.14 蛟河草莓研究所
Jiaohe Strawberry Research Institute

2018年6月李怀宝高级经济师在甘肃礼县考察野生草莓资源
Senior economist Huaibao Li investigated the wild strawberry resources in Lixian County, Gansu Province in June, 2018.

1.15 重庆元邦农业发展有限公司
Chongqing Yuanbang Agricultural Development Co., Ltd.

2008年7月秦国新博士在山西五台山考察野生草莓
Dr. Guoxin Qin investigated the wild strawberry resources in Mount Wutai, Shanxi Province in July, 2008.

1.16 中国园艺学会草莓分会
Strawberry Section, Chinese Society for Horticultural Science

2016年5月在四川成都考察野生草莓
The wild strawberry resources investigation in Chengdu City, Sichuan Province in May, 2016.

2016年6月在西藏林芝考察野生草莓
The wild strawberry resources investigation in Nyingchi County, Xizang in June, 2016.

2017年6月在四川峨眉山考察野生草莓
The wild strawberry resources investigation in Mount Emei, Sichuan Province in June, 2017.

2017年6月在云南迪庆考察野生草莓
The wild strawberry resources investigation in Diqing Prefecture, Yunnan Province in June, 2017.

2018年5月在贵州梵净山考察野生草莓
The wild strawberry resources investigation in Mount Fanjing, Guizhou Province in May, 2018.

2018年6月在甘肃天水考察野生草莓
The wild strawberry resources investigation in Tianshui City, Gansu Province in June, 2018.

2018年6月在甘肃礼县考察野生草莓
The wild strawberry resources investigation in Lixian County, Gansu Province in June, 2018.

2019年6月在黑龙江尚志考察野生草莓
The wild strawberry resources investigation in Shangzhi City, Heilongjiang Province in June, 2019.

2021年6月在湖北神农架考察野生草莓
The wild strawberry resources investigation in Shennongjia Forest Region, Hubei Province in June, 2021.

2021年7月在吉林长白山考察野生草莓
The wild strawberry resources investigation in Mount Changbai, Jilin Province in July, 2021.

2023年5月在陕西略阳考察野生草莓
The wild strawberry resources investigation in Lueyang County, Shaanxi Province in May, 2023.

2023年5月在陕西宁强考察野生草莓
The wild strawberry resources investigation in Ningqiang County, Shaanxi Province in May, 2023.

2023年6月在湖北神农架考察野生草莓
The wild strawberry resources investigation in Shennongjia Forest Region, Hubei Province in June, 2023.

2023年6月在湖北利川考察野生草莓
The wild strawberry resources investigation in Lichuan City, Hubei Province in June, 2023.

2024年7月在山西沁源考察野生草莓
The wild strawberry resources investigation in Qinyuan County, Shanxi Province in July, 2024.

第二章
Chapter 2

中国野生草莓资源的保存
Conservation of wild strawberry resources in China

我国一些科研单位、公司、私人爱好者等均开展了对我国野生草莓资源的收集保存工作。目前，以沈阳农业大学、北京市农林科学院林业果树研究所和江苏省农业科学院果树研究所保存的野生草莓资源较多，均收集保存有400～900份资源，其他科研单位也保存了相当数量的野生草莓资源，为开展我国野生草莓资源的研究和利用奠定了良好基础。

目前，除了在自然界中原生境就地保存外，另一种主要保存方式是迁地保存。迁地保存中，一些单位以露地栽植保存为主（如沈阳农业大学、北京市农林科学院、中国农业科学院郑州果树研究所、黑龙江省农业科学院园艺分院、西北农林科技大学、贵州省农业科学院园艺所、昆明市农业科学院、蛟河草莓研究所等），也有的单位以温室大棚盆栽保存为主（如江苏省农业科学院、湖北省农业科学院经作所、四川省农业科学院园艺所、杭州市农业科学院生物技术研究所、云南农业大学、甘肃省农业科学院林果花卉所等）。目前，我国开展野生草莓资源的种子、茎尖离体培养、超低温等保存方式的研究和应用还很少。

Some scientific research institutes, companies and private enthusiasts have collected and preserved wild strawberry resources in China. At present, Shenyang Agricultural University, Institute of Forestry and Fruit Trees of Beijing Academy of Agriculture and Forestry Sciences, and Jiangsu Academy of Agricultural Sciences have preserved 400~900 accessions of wild strawberry resources, and other scientific research institutes have also preserved a considerable amount of wild strawberry resources, which has laid a good foundation for the research and utilization of wild strawberry resources native to China.

At present, in addition to in situ preservation, another main preservation method is ex situ preservation. Some institutes preserved the wild strawberry resources using planting in the open field, such as Shenyang Agricultural University, Beijing Academy of Agriculture and Forestry Sciences, Zhengzhou Fruit Research Institute of Chinese Academy of Sciences, Horticulture Branch of Heilongjiang Academy of Agricultural Sciences, Northwest A&F University, Horticulture Institute of Guizhou Academy of Agricultural Sciences, Kunming Academy of Agricultural Sciences, and Jiaohe Strawberry Research Institute. Some institutes preserved the wild strawberry resources using pot in greenhouse, such as Jiangsu Academy of Agricultural Sciences, Industrial Crops Institute, Hubei Academy of Agricultural Sciences, Horticulture Institute of Sichuan Academy of Agricultural Sciences, Institute of Biotechnology, Hangzhou Academy of Agricultural Sciences, Yunnan Agricultural University, and Fruit and Floriculture Research Institute of Gansu Academy of Agricultural Sciences. In China, there are few researches on seed preservation, meristem culture preservation, and cryopreservation of wild strawberry resources.

2.1 沈阳农业大学

Shenyang Agricultural University

沈阳农业大学是世界上收集保存草莓属植物种类最全、份数最多的研究单位之一，系统开展了野生草莓源的收集、分类、保存、评价、种间杂交、倍性育种等方面的研究。目前共收集野生草莓资源966份，保存有全世界草莓属植物23个野生种，包括五叶草莓、黄毛草莓、森林草莓、西藏草莓、裂萼草莓、绿色草莓、东北草莓、中国草莓、台湾草莓、峨眉草莓、日本草莓、饭沼草莓、两季草莓、东方草莓、西南草莓、伞房草莓、纤细草莓、高原草莓、麝香草莓、智利草莓、弗州草莓、择捉草莓、喀斯喀特草莓等。沈阳农业大学雷家军教授在国内首次赋予了7个野生草莓种的中文名称，包括东北草莓、高原草莓、饭沼草莓、择捉草莓、两季草莓、布哈拉草莓、喀斯喀特草莓，并于2021年鉴定命名了草莓属植物一个新种峨眉草莓（*Fragaria emeiensis* Jia J. Lei）。

Shenyang Agricultural University is one of the research institutes collecting and preserving wild strawberry resources in the world, and has focused on the collection, taxonomy, conservation, evaluation, interspecific hybridization and ploidy breeding of wild strawberry resources. Up to now, a total of 966 wild strawberry accessions involved in 23 wild species have been collected and preserved, including *F. pentaphylla, F. nilgerrensis, F. vesca, F. nubicola, F. daltoniana, F. viridis, F. mandschurica, F. chinensis, F. hayatai, F. emeiensis, F. nipponica, F. iinumae, F. bifera, F. orientalis, F. moupinensis, F. corymbosa, F. gracilis, F. tibetica, F. moschata, F. chiloensis, F. virginiana, F. iturupensis,* and *F. cascadensis.* Prof. Jiajun Lei firstly gave the Chinese names of 7 wild strawberry species in China, including *F. mandschurica, F. tibetica, F. iinumae, F. iturupensis, F. bifera, F. bucharica,* and *F. cascadensis,* and identified a new species (*F. emeiensis* Jia J. Lei) in the genus *Fragaria* in 2021.

沈阳农业大学保存的野生草莓资源（2013年）
Wild strawberry resources preserved in Shenyang Agricultural University in 2013

沈阳农业大学保存的野生草莓资源（2019年）
Wild strawberry resources preserved in Shenyang Agricultural University in 2019

沈阳农业大学保存的野生草莓资源（2023年）
Wild strawberry resources preserved in Shenyang Agricultural University in 2023

2.2 北京市农林科学院林业果树研究所

Forestry and Fruit Research Institute, Beijing Academy of Agriculture and Forestry Sciences

近年来考察了吉林长白山地区；四川省广元市、巴中市；陕西省汉中市；西藏自治区林芝市；云南迪庆州、楚雄州、曲靖市、怒江州；青海海东市；甘肃天水市、陇南市、定西市、临夏州；黑龙江省的尚志市、牡丹江市、伊春市、齐齐哈尔市；湖北省神农架林区、恩施州；河北省张家口市、保定市等地。通过不同途径收集了20个种400余份野生草莓资源，包括森林草莓、黄毛草莓、绿色草莓、裂萼草莓、五叶草莓、东北草莓、中国草莓、日本草莓、饭沼草莓、东方草莓、西南草莓、伞房草莓、麝香草莓、智利草莓等。对收集到的野生资源进行精准鉴定，获得了耐盐种质绿色草莓、香味种质黄毛草莓、高可溶性固形物含量东方草莓等，为草莓香气育种、抗性育种及相关理论研究提供了重要的资源。开展了基因组、代谢组等多组学研究，基于叶绿体基因组测序开展了草莓属进化研究，推测草莓属野生资源起源于青藏高原地区，草莓属现存的最古老的种为饭沼草莓，青藏高原隆起所带来的区域气候变化和300万年以来全球气候变化可能是草莓属不同类群形成和倍性提高的主要推动力。完成了五叶草莓、绿色草莓等资源的全基因组测序，获得了高质量的基因组。比较基因组分析发现，五叶草莓与西藏草莓最为接近。

In recent years, many sites have been investigated, including: the Changbai Mountain area in Jilin Province; Guangyuan City and Bazhong City in Sichuan Province; Hanzhong City in Shaanxi Province; Nyingchi City in Xizang Autonomous Region; Diqing Prefecture, Chuxiong Prefecture, Qujing City and Nujiang Prefecture in Yunnan Province; Haidong City in Qinghai Province; Tianshui City, Longnan City, Dingxi City and Linxia Prefecture in Gansu Province; Shangzhi City, Mudanjiang City, Yichun City and Qiqihar City in Heilongjiang Province; Shennongjia Forest Region and Enshi Prefecture in Hubei Province; Zhangjiakou City and Baoding City in Hebei Province. More than 400 wild strawberry resources involved in 20 species were collected through different ways, including *F. vesca*, *F. nilgerrensis*, *F. viridis*, *F. daltoniana*, *F. pentaphylla*, *F. mandschurica*, *F. chinensis*, *F. nipponica*, *F. iinumae*, *F. orientalis*, *F. moupinensis*, *F. corymbosa*, *F. moschata*, and *F. chiloensis*. Some wild resources were accurately identified. The germplasms of salt-tolerant *F. viridis*, fragrant *F. nilgerrensis*, and *F. orientalis* with high soluble solid content were obtained, which provided important resources for aroma breeding, resistance breeding and related theoretical research of strawberry. Based on chloroplast genome sequencing, the evolution of strawberry has been studied. It is speculated that the wild resources of strawberry originated from the Qinghai-Xizang Plateau, and the oldest existing species of strawberry is *F. iinumae*. The regional climate change caused by the Xizang Plateau rise and the global climate change in the past three million years may be the main driving force for the formation of different strawberry populations and the increase of ploidy. The whole genome sequencing of *F. pentaphylla* and *F. viridis* was completed, and high-quality genome was obtained. Comparative genomic analysis showed that *F. pentaphylla* was the closest to *F. nipponica*.

中国野生草莓资源
Wild Strawberry Resources in China

北京市农林科学院林业果树研究所的国家北京草莓资源圃
National Strawberry Germplasm Repository in Beijing located in Forestry and Fruit Research Institute, Beijing Academy of Agriculture and Forestry Sciences

北京市农林科学院林业果树研究所保存的野生草莓资源
The wild strawberry resources preserved in Forestry and Fruit Research Institute, Beijing Academy of Agriculture and Forestry Sciences

2.3 江苏省农业科学院果树研究所
Fruit Research Institute, Jiangsu Academy of Agricultural Sciences

江苏省农业科学院于1948年开展草莓资源引种工作，1981年启动建设国家果树种质南京桃、草莓资源圃。持续开展野生资源考察收集工作，足迹遍布全国野生草莓分布区的西藏自治区林芝市，云南省迪庆州、丽江市、曲靖市，贵州省毕节市、梵净山，四川省乐山市、广元市，湖北省神农架、恩施，甘肃省临夏市、陇南市，陕西省汉中市，山西忻州市，青海省西宁市、海东市，新疆乌鲁木齐市，黑龙江省哈尔滨市、伊春市、牡丹江市，吉林省白山市等。现已收集保存国内外各类野生草莓资源778份，涉及全世界23个野生种，包括12个二倍体种、5个四倍体种、1个自然五倍体种、1个六倍体种、2个八倍体种和2个十倍体种。开展了野生资源的鉴定评价与种间杂交工作，筛选出特异香气、特异果色、果实性状、雌雄蕊变异、抗性等性状相关的优异野生种质40份。基于二倍体草莓的SSR分子标记，构建了核心种质库。明确中国14个野生种地域分布特征，首次阐明栽培种草莓的祖先种，明确二倍体野生种质的系统进化和亲缘关系。编著了《草莓种质资源描述规范和数据标准》、制定及参与制定了《草莓种质资源描述规范》《农作物优异种质资源评价规范 草莓》《农作物种质资源鉴定技术规程 草莓》等行业标准3项，广为国内同行应用。

Jiangsu Academy of Agricultural Sciences started the collection of strawberry resources in 1948, and the National Fruit Germplasm Nanjing Peach and Strawberry Resource Repository in 1981. It has continuously carried out the investigation and collection of wild resources, covering Nyingchi City in Xizang Autonomous Region; Deqing Prefecture, Lijiang City and Qujing City in Yunnan Province; Bijie City and Mount Fanjing in Guizhou Province; Leshan City and Guangyuan City in Sichuan Province; Shennongjia Forest Region and Enshi City in Hubei Province; Linxia City and Longnan City in Gansu Province; Hanzhong City in Shaanxi Province; Xinzhou City in Shanxi Province; Xining City and Haidong City in Qinghai Province; Urumqi City in Xinjiang; Harbin City, Yichun City and Mudanjiang City in Heilongjiang Province; Baishan City in Jilin Province, etc. At present, 778 wild strawberry resources at home and abroad have been collected and preserved, involving 23 wild species in the world, including 12 diploid species, 5 tetraploid species, 1 natural pentaploid species, 1 hexaploid species, 2 octoploid species and 2 decaploid species. The identification and evaluation of wild resources and interspecific hybridization were carried out, and 40 excellent wild germplasm related to specific aroma, specific fruit color, fruit traits, female and stamen variation, resistance and other traits were selected. A core germplasm bank was constructed based on SSR markers of diploid strawberry. The geographical distribution characteristics of 14 wild species in China was studied, the ancestor species of cultivated strawberry was elucidated, and the phylogenetic evolution and relationship of diploid wild species were clarified. *The Description Specification and Data Standard of Strawberry Germplasm Resources*, and three industry standards including *The Description Standard of Strawberry Germplasm Resources*, *The Specification for Evaluation of Excellent Crop Germplasm Resources: Strawberry*, and *The Technical Specification for Identification of Crop Germplasm Resources: Strawberry*, were published and widely applied by peers in China.

江苏省农业科学院的国家南京草莓资源圃
National Strawberry Germplasm Repository in Nanjing located in Jiangsu Academy of Agricultural Sciences

江苏省农业科学院果树研究所保存的野生草莓资源
The wild strawberry resources preserved in Fruit Research Institute, Jiangsu Academy of Agricultural Sciences

2.4 中国农业科学院郑州果树研究所

Zhengzhou Fruit Research Institute, China Academy of Agricultural Sciences

中国农业科学院郑州果树研究所是国家园艺种质资源库草莓分库（郑州）的依托单位。多年来考察收集了我国川藏高山峡谷区318国道沿线、藏东南林芝、波密、墨脱等地，以及云南、贵州、四川、甘肃、新疆、陕西、黑龙江等地的野生草莓资源。现保存包括黄毛草莓、森林草莓、五叶草莓、绿色草莓、西藏草莓、东北草莓、裂萼草莓、中国草莓、峨眉草莓、饭沼草莓、日本草莓、两季草莓、东方草莓、西南草莓、伞房草莓、纤细草莓、高原草莓、麝香草莓、智利草莓、弗州草莓、择捉草莓等21个野生种、455份资源。对收集保存的野生种质进行了较为系统的调查和鉴定评价，筛选抗病、耐盐碱种质，挖掘抗性基因；通过对21个种叶绿体基因组进行比较基因组学分析，揭示草莓种群体进化。

Zhengzhou Fruit Research Institute of Chinese Academy of Agricultural Sciences is the supporting organization of Strawberry Branch Bank of National Horticultural Germplasm Resource Bank (Zhengzhou). Over the years, wild strawberry resources have been collected from mountains and valleys along the 318 National Highway in Sichuan and Xizang, Nyingchi County, Bomi County, and Motuo County in southeast Xizang, as well as Yunnan, Guizhou, Sichuan, Gansu, Xinjiang, Shaanxi and Heilongjiang Provinces. The 455 wild strawberry accessions involved in 21 species were collected, including *F. nilgerrensis*, *F. vesca*, *F. pentaphylla*, *F. viridis*, *F. nubicola*, *F. mandschurica*, *F. daltoniana*, *F. chinensis*, *F. emeiensis*, *F. iinumae*, *F. nipponica*, *F. orientalis*, *F. moupinensis*, *F. corymbosa*, *F. gracilis*, *F. tibetica*, *F. moschata*, *F. chiloensis*, *F. virginiana*, *F. iturupensis*, and *F. cascadensis*. The wild germplasms were collected, preserved, investigated and evaluated systematically. The germplasms resistant to disease and saline-alkali were screened and the resistance genes were mined. Through comparative genomic analysis of chloroplast genome from 21 *Fragaria* species, the strawberry evolution was further revealed.

中国农业科学院郑州果树研究所保存的野生草莓资源
The wild strawberry resources preserved in Zhengzhou Fruit Research Institute, China Academy of Agricultural Sciences

2.5 湖北省农业科学院经济作物所
Industrial Crops Institute, Hubei Academy of Agricultural Sciences

湖北省农业科学院经济作物所近年来考察了湖北省的神农架、巴东、利川、长阳、秭归，湖南省的石门，陕西省的略阳、宁强，甘肃省的康县等地的野生草莓资源。通过各种途径收集了13个种95份野生草莓资源，包括黄毛草莓、东北草莓、东方草莓、绿色草莓、森林草莓、中国草莓、五叶草莓、西藏草莓、伞房草莓、西南草莓、麝香草莓、弗州草莓、智利草莓等。开展了对野生资源的抗病性、品质评价及与栽培品种远缘杂交等方面工作。

In recent years, the wild strawberry resources in Shennongjia Forest Region, Badong County, Lichuan City, Changyang County and Zigui County in Hubei Province, Shimen in Hunan Province, Lueyang County and Ningqiang County in Shaanxi Province, and Kang County in Gansu Province were investigated. The 95 wild strawberry accessions involved in 13 species were collected through various ways, including *F. nilgerrensis*, *F. mandschurica*, *F. orientalis*, *F. viridis*, *F. vesca*, *F. chinensis*, *F. pentaphylla*, *F. nubicola*, *F. corymbosa*, *F. moupinensis*, *F. moschata*, *F. virginiana*, and *F. chiloensis*. The researches on disease resistance, quality evaluation and distant hybridization of wild strawberry resources were carried out.

湖北省农业科学院经济作物所保存的野生草莓资源
The wild strawberry resources preserved in Industrial Crops Institute, Hubei Academy of Agricultural Sciences

2.6 四川省农业科学院园艺研究所

Horticulture Research Institute, Sichuan Academy of Agricultural Sciences

　　2004年至今，四川省农业科学院园艺所对四川盆周山区、川西高原—川西南高海拔区—青东南—藏东—藏东南—滇西北、秦巴山区（川北—川东北、甘东南、陕南）、乌蒙山区（川南、滇东北、黔西南）、渝东南、黔东北，以及天山山脉等我国西南地区及部分西北地区进行了大量考察。目前保存国内外野生草莓17个种485份资源，包括黄毛草莓、五叶草莓、森林草莓、峨眉草莓、西藏草莓、绿色草莓、中国草莓、台湾草莓、日本草莓、东北草莓、东方草莓、西南草莓、伞房草莓、纤细草莓、麝香草莓、智利草莓、弗州草莓等。已开展"四川及其周边地区野生草莓资源遗传多样性及亲缘关系研究""不同地理来源的黄毛草莓果实香气成分初步分析"，同时正利用野生草莓优异资源与优良栽培品种进行种内或种间育种创新研究。

　　From 2004 to now, Horticultural Research Institute of Sichuan Academy of Agricultural Sciences has investigated the Penzhou Mountain area in Sichuan, the western Sichuan Plateau–southwest Sichuan high altitude area–southeast Qinghai–east Xizang–southeast Xizang–northwest Yunnan, Qinba Mountain area (north Sichuan–northeast Sichuan, southeast Gansu, and south Shaanxi), Wumeng Mountain area (south Sichuan, northeast Yunnan, and southwest Guizhou), southeast Chongqing, and northeast Guizhou. A lot of investigations were carried out in Tianshan Mountains and southwest and northwest China. The 485 accessions of wild strawberry involved in 17 species were collected and preserved, including *F. nilgerrensis*, *F. pentaphylla*, *F. vesca*, *F. emeiensis*, *F. nubicola*, *F. viridis*, *F. chinensis*, *F. hayatai*, *F. nipponica*, *F. mandschurica*, *F. orientalis*, *F. moupinensis*, *F. corymbosa*, *F. gracilis*, *F. moschata*, *F. chiloensis*, and *F. virginiana*. We have researched on genetic diversity and phylogenetic relationship of wild strawberry resources in Sichuan and its surrounding areas, and on preliminary analysis of aroma components of *F. nilgerrensis* fruits from different geographical sources. The intraspecific or interspecific hybridization is being carried out using excellent wild strawberry resources and cultivars.

四川省农业科学院园艺研究所保存的野生草莓资源
The wild strawberry resources preserved in Horticulture Research Institute, Sichuan Academy of Agricultural Sciences

2.7 云南农业大学

Yunnan Agricultural University

云南农业大学乔琴教授与中国科学院昆明植物研究所张体操研究员合作，近年来对云南、西藏、四川、甘肃、陕西等地等的野生草莓资源进行了多次考察，并通过从沈阳农业大学野生草莓资源圃引种，目前收集野生草莓资源15个种180份，包括黄毛草莓、西南草莓、五叶草莓、中国草莓、饭沼草莓、西藏草莓、高原草莓、裂萼草莓、森林草莓、东北草莓、绿色草莓、伞房草莓、东方草莓、弗州草莓、智利草莓等。利用这些野生草莓资源开展了多倍体诱导、原生质体融合、远缘杂交等种质创新及重要性状基因挖掘等方面的工作。

In recent years, Prof. Qin Qiao worked in Yunnan Agricultural University and Prof. Ticao Zhang worked in Kunming Institute of Botany, Chinese Academy of Sciences have conducted a lot of investigations on wild strawberry resources in Yunnan, Xizang, Sichuan, Gansu, and Shaanxi Provinces. At present, 180 wild strawberry accessions of 15 species have been collected, including *F. nilgerrensis*, *F. moupinensis*, *F. pentaphylla*, *F. chinensis*, *F. iinumae*, *F. nubicola*, *F. tibetica*, *F. daltoniana*, *F. vesca*, *F. mandschurica*, *F. viridis*, *F. corymbosa*, *F. orientalis*, *F. virginiana*, and *F. chiloensis*. These wild strawberry resources were used for germplasm innovation, such as polyploid induction, protoplast fusion, distant hybridization, and gene mining of important traits.

乔琴教授和张体操研究员收集保存的野生草莓资源
The wild strawberry resources collected and preserved by Prof. Qin Qiao and Prof. Ticao Zhang

2.8 杭州市农业科学院生物技术研究所
Institute of Biotechnology, Hangzhou Academy of Agricultural Sciences

　　近年来，杭州市农业科学院重点对陕西、甘肃、四川、云南、贵州、湖北、山西、青海等地野生草莓资源进行了考察，目前收集野生草莓资源8个种70余份，保存有黄毛草莓、五叶草莓、绿色草莓、中国草莓、森林草莓、西南草莓、麝香草莓、弗州草莓等。利用这些野生草莓资源开展了抗性调查、果实品质评价分析等方面的工作，希望将我国优良的野生草莓应用在草莓生产上。

　　In recent years, Hangzhou Academy of Agricultural Sciences has carried out some investigations of wild strawberry resources in Shaanxi, Gansu, Sichuan, Yunnan, Guizhou, Hubei, Shanxi and Qinghai Provinces. At present, more than 70 wild strawberry accessions involved in 8 *Fragaria* species have been collected, including *F. nilgerrensis*, *F. pentaphylla*, *F. viridis*, *F. chinensis*, *F. vesca*, *F. moupinensis*, *F. moschata*, and *F. virginiana*. The resistance investigation, evaluation and analysis of fruit quality of these wild strawberry resources were carried out in order to utilize them in commercial production in the future.

杭州市农业科学院生物技术研究所保存的野生草莓和品种资源
The wild strawberry resources and cultivars preserved in Institute of Biotechnology, Hangzhou Academy of Agricultural Sciences

2.9 黑龙江省农业科学院园艺分院
Horticultural Branch, Heilongjiang Academy of Agricultural Sciences

黑龙江省农业科学院园艺分院近年来考察了黑龙江省的尚志市、牡丹江市、伊春市、大兴安岭，吉林长白山，陕西秦岭等的野生草莓资源。通过不同途径收集了10个种56份野生草莓资源，包括东北草莓、东方草莓、森林草莓、西南草莓、中国草莓、饭沼草莓、黄毛草莓、五叶草莓、绿色草莓、麝香草莓等。黑龙江省农业科学院园艺分院正在通过诱变育种、远缘杂交选育等对野生资源进行利用。

In recent years, Horticultural Branch of Heilongjiang Academy of Agricultural Sciences has investigated the wild strawberry resources in Shangzhi City, Mudanjiang City and Greater Hinggan Mountains in Heilongjiang Province, Mount Changbai in Jilin Province, and Qinling Mountains in Shaanxi Province. The 56 wild strawberry accessions involved in 10 species were collected, including *F. mandschurica*, *F. orientalis*, *F. vesca*, *F. moupinensis*, *F. chinensis*, *F. iinumae*, *F. nilgerrensis*, *F. pentaphylla*, *F. viridis*, and *F. moschata*. Horticultural Branch of Heilongjiang Academy of Agricultural Sciences is making use of wild resources through mutation breeding and distant hybridization.

黑龙江省农业科学院园艺分院保存的野生草莓资源
The wild strawberry resources preserved in Horticultural Branch, Heilongjiang Academy of Agricultural Sciences

2.10 西北农林科技大学
Northwest A&F University

西北农林科技大学从2018年开始对野生草莓资源进行考察和收集，主要对陕西省的略阳、宁强、太白、陇县，甘肃省的康县、徽县、宕昌县、岷县，四川省的平武县，云南省的寻甸县，湖北省的神农架、巴东、利川等地的野生草莓资源进行考察收集，并在陕西省略阳县建立了秦巴山区野生草莓种质资源圃。目前已经收集了5个种135份野生草莓资源，包括五叶草莓、黄毛草莓、森林草莓、中国草莓、麝香草莓。开展了对野生资源的抗病性、果型、果色、香气等品质方面的评价工作，并对五叶草莓进行了化学诱变育种，产生了大量不同类型的突变体。对略阳县野生草莓产业开发利用进行了指导和宣传。

Since 2018, Northwest A&F University has been carrying out a lot of investigation and collection of wild strawberry resources in Lueyang County, Ningqiang County, Taibai County and Long County in Shaanxi Province, Kang County, Hui County and Min County in Gansu Province, Pingwu County in Sichuan Province, Xundian County in Yunnan Province, and Shennongjia Forest Region, Badong County and Lichuan City in Hubei Province. A germplasm repository of wild strawberry resources was established in Lueyang County located in Qinba Mountain area, Shaanxi Province. At present, 135 wild strawberry accessions of 5 species have been collected and preserved, including *F. pentaphylla*, *F. nilgerrensis*, *F. vesca*, *F. chinensis*, and *F. moschata*. The evaluation of disease resistance, fruit type, fruit color, aroma and other qualities of wild resources was estimated, and chemical mutation breeding of *F. pentaphylla* was carried out, and a lot of mutants were obtained. We guided and publicized the development and utilization of wild strawberry industry in Lueyang County.

西北农林科技大学在秦巴山区保存的野生草莓资源
The wild strawberry resources preserved in the Qinba Mountain area by Northwest A&F University

2.11 贵州省农业科学院园艺研究所
Institute of Horticulture, Guizhou Academy of Agricultural Sciences

贵州省农业科学院园艺研究所草莓课题组自2013年开始，陆续在贵州境内开展野生草莓资源的考察与收集工作。足迹遍布贵阳市花溪区、白云区、乌当区，安顺市西秀区、镇宁县、紫云县，毕节市七星关区、赫章县、威宁县，遵义市赤水市、仁怀市、习水县，铜仁市江口县，六盘水市钟山区、盘州市，黔东南州凯里市、雷山县、麻江县，黔南州都匀市、龙里县、惠水县、长顺县，黔西南州望谟县等地。共收集到4个种31份野生草莓资源，包括黄毛草莓、五叶草莓、西南草莓、中国草莓等。近年来，利用黄毛草莓作为亲本，通过远缘杂交选育出了'黔莓六号''黔莓七号'等新品种（系）。

Since 2013, the strawberry research group from Institute of Horticulture, Guizhou Academy of Agricultural Sciences has carried out the investigation and collection of wild strawberry resources in Guizhou Province. It has covered Huaxi District, Baiyun District and Wudang District in Guiyang City; Xixiu District, Zhenning County and Ziyun County in Anshun City; Qixingguan District, Hezhang County and Weining County in Bijie City; Chishui City, Renhuai City and Xishui County in Zunyi City; Jiangkou County in Tongren City; Zhongshan District and Panzhou City in Liupanshui City; Kaili City, Leishan County and Majiang County in Qiandongnan Prefecture; Duyun City, Longli County, Huishui County and Changshun County in Qiannan Prefecture; Wangmo County in Qianxinan Prefecture, etc. A total of 31 wild strawberry accessions involved in 4 species were collected, including *F. nilgerrensis*, *F. pentaphylla*, *F. moupinensis*, and *F. chinensis*. In recent years, several new cultivars such as 'Qianmei No. 6' and 'Qianmei No. 7' have been released by distant hybridization using *F. nilgerrensis* as a parent.

贵州省农业科学院园艺研究所保存的野生草莓资源
The wild strawberry resources preserved in Institute of Horticulture, Guizhou Academy of Agricultural Sciences

2.12 甘肃省农业科学院林果花卉研究所
Fruit and Floriculture Research Institute, Gansu Academy of Agricultural Sciences

自2012年以来，甘肃省农业科学院林果花卉研究所考察了甘肃省兰州市、天水市、陇南市、甘南县等地的野生草莓资源。目前收集保存了5个种28份野生草莓资源，包括森林草莓、五叶草莓、伞房草莓、西南草莓、东方草莓等。目前，正在开展野生草莓和栽培品种的辐射诱变和航天育种工作，先后获得诱变种子3000余粒，诱变植株216株，为下一步草莓种质创新、新品种选育奠定了基础。

Since 2012, Fruit and Floriculture Research Institute of Gansu Academy of Agricultural Sciences has investigated the wild strawberry resources in Lanzhou City, Tianshui City, Longnan City and Gannan Prefecture in Gansu Province. At present, 28 wild strawberry resources of 5 species have been collected and preserved, including *F. vesca*, *F. pentaphylla*, *F. corymbosa*, *F. moupinensis*, and *F. orientalis*. At present, radiation mutation and space breeding of wild strawberry and cultivated strawberry cultivars are being carried out, and more than 3,000 mutation seeds and 216 mutation plants have been obtained, which has laid the foundation for the next step of strawberry germplasm innovation and new cultivar breding.

甘肃省农业科学院林果花卉研究所保存的野生草莓资源
The wild strawberry resources preserved in Fruit and Floriculture Research Institute, Gansu Academy of Agricultural Sciences

2.13 昆明市农业科学院
Kunming Academy of Agricultural Sciences

昆明市农业科学院近年来考察了云南省昆明市、丽江市、玉溪市、曲靖市、临沧市、保山市、普洱市、楚雄州、红河州、文山州、西双版纳州、德宏州、迪庆州、大理州、怒江州等15个州（市）52个县（区）的野生草莓资源，收集了4个种84份草莓野生资源，主要包括西南草莓、黄毛草莓、五叶草莓、西藏草莓等，收集的资源最高海拔为4 323 m。与云南大学微生物研究所合作开展了"云南草莓属植物的微生物组生物地理特征研究"，并尝试利用野生资源通过诱变育种和杂交育种获得优良材料和新品种。

In recent years, Kunming Academy of Agricultural Sciences has investigated the wild strawberry resources in 52 counties or districts in 15 prefectures or cities including Kunming City, Lijiang City, Yuxi City, Qujing City, Lincang City, Baoshan City, Puer City, Chuxiong Prefecture, Honghe Prefecture, Wenshan Prefecture, Xishuangbanna Prefecture, Dehong Prefecture, Diqing Prefecture, Dali Prefecture and Nujiang Prefecture in Yunnan Province, and 84 wild strawberry accessions of 4 species were collected, including *F. moupinensis*, *F. nilgerrensis*, *F. pentaphylla*, and *F. nubicola*. The highest altitude of the collected resources is up to 4,323 m. In cooperation with Microbiology Institute of Yunnan University, we carried out study on the biogeographic characteristics of microbiome of wild strawberry in Yunnan Province, and tried to use wild strawberry resources to obtain excellent materials and new cultivars through mutation breeding and distant hybridization.

昆明市农业科学院保存的野生草莓资源
The wild strawberry resources preserved in Kunming Academy of Agricultural Sciences

2.14 蛟河草莓研究所
Jiaohe Strawberry Research Institute

　　蛟河草莓研究所是一家专业从事草莓品种、技术、示范、推广的民营企业。在北方寒地建有一处草莓种质资源圃。从2001年至今，对吉林、云南、贵州、四川、甘肃、宁夏、青海、西藏、黑龙江、湖北、陕西等省进行了野生草莓资源考察收集，现保存野生草莓资源8个种123份，收集保存种类包括东北草莓、东方草莓、森林草莓、中国草莓、黄毛草莓、五叶草莓、西南草莓、弗州草莓等。开展了野生草莓资源的抗寒性试验、杂交利用等工作。

Jiaohe Strawberry Research Institute is a private enterprise specializing in strawberry cultivars, technology, demonstration and popularization. A strawberry germplasm resource repository was established in the cold region of northeast China. Since 2001, wild strawberry resources have been investigated and collected in Jilin, Yunnan, Guizhou, Sichuan, Gansu, Ningxia, Qinghai, Xizang, Heilongjiang, Hubei, and Shaanxi Provinces, and 123 wild strawberry accessions of 8 species have been preserved, including *F. mandschurica*, *F. orientalis*, *F. vesca*, *F. chinensis*, *F. nilgerrensis*, *F. pentaphylla*, *F. moupinensis*, and *F. virginiana*. The cold resistance and cross utilization of wild strawberry resources were carried out.

蛟河草莓研究所保存的野生草莓资源
The wild strawberry resources preserved in Jiaohe Strawberry Research Institute

2.15 重庆元邦农业发展有限公司
Chongqing Yuanbang Agricultural Development Co., Ltd.

从2000年以来，该公司秦国新博士一直进行野生草莓种质资源调查、收集工作。重点对我国西南地区进行了调查，先后考察了重庆、贵州、陕西、四川、山西、云南等地，收集了中国原产7个野生草莓种类60份资源，包括黄毛草莓、五叶草莓、西藏草莓、峨眉草莓、纤细草莓、伞房草莓、西南草莓等。进行了中国草莓属植物遗传多样性及亲缘演化关系的研究，建立了草莓属植物SCoT分析体系，并与沈阳农业大学、日本香川大学等单位共同完成了部分草莓属植物根尖染色体的GISH分析。

Dr. Guoxin Qin has investigated and collected wild strawberry resources since 2000. Some investigations were focused in southwest China, including Chongqing, Guizhou, Shaanxi, Sichuan, Shanxi, and Yunnan Provinces. Sixty accessions involved in 7 wild strawberry species distributed in China were collected, including *F. nilgerrensis*, *F. pentaphylla*, *F. nubicola*, *F. emeiensis*, *F. gracilis*, *F. corymbosa*, and *F. moupinensis*. The genetic diversity and phylogenetic relationship of *Fragaria* species native to China were studied. The SCoT analysis of the genus *Fragaria* was established, and GISH analysis of root tip chromosome of some *Fragaria* species was completed jointly with Shenyang Agricultural University in China and Kagawa University in Japan.

秦国新博士收集保存的部分野生草莓资源
The wild strawberry resources collected and preserved by Dr. Guoxin Qin

中国野生草莓种类的分布
Distribution of wild strawberry species in China

3.1 中国野生草莓种类分布概况
Distribution outline of wild strawberry species in China

中国草莓属植物自然分布有15个种，包括10个二倍体种和5个四倍体种（见第四章）。中国野生草莓资源丰富，分布非常广泛。中国的天山、长白山、秦岭、大兴安岭、青藏高原、云贵高原是天然的野生草莓基因库，蕴藏着种类和数量丰富的野生草莓，存在较多的种、变种和类型，其中许多为特有珍稀的优良资源。

There are 15 wild *Fragaria* species distributed in China, including 10 diploid species and 5 tetraploid species (see Chapter 4). China is rich in wild strawberry resources, which distribute widely. Wild strawberry populations are extensively distributed in Tianshan Mountains, Changbai Mountains, Qinling Mountains, Daxing'an Mountains, Qinghai-Xizang Plateau Highland and Yunnan-Guizhou Plateau in China, where some special and rare *Fragaria* species, variations and types were discovered.

中国野生草莓资源有4个集中分布区：第一个是中国西南部地区，主要包括四川、云南、贵州、重庆、西藏及陕西南部，这个区分布的野生草莓种类最多，约8个种，包括黄毛草莓、五叶草莓、西藏草莓、裂萼草莓、峨眉草莓、中国草莓、西南草莓、高原草莓；第二个是中国西北部和中部地区，主要包括甘肃、青海、陕西、河北、山西、河南等省，这个区分布的野生草莓种类也较多，约5个种，包括五叶草莓、中国草莓、黄毛草莓、伞房草莓、纤细草莓；第三个是中国东北地区，主要包括吉林、黑龙江、辽宁、内蒙古，分布有3个种，包括东方草莓、东北草莓、森林草莓；第四个是中国新疆地区，分布有2个种，为绿色草莓、森林草莓。

There are 4 concentrated distribution areas of wild strawberry resources in China. The first one is southwest China, which mainly includes Sichuan Province, Yunnan Province, Guizhou Province, Chongqing City, Xizang Autonomous Region and southern Shaanxi Province. There are the most wild strawberry species distributed in this area, involving about 8 species including *F. nilgerrensis*, *F. pentaphylla*, *F. nubicola*, *F. daltoniana*, *F. emeiensis*, *F. chinensis*, *F. moupinensis*, and *F. Tibetica*. The second is the northwest and central regions of China, including Gansu Province, Qinghai Province, Shaanxi Province, Hebei Province, Shanxi Province, and Henan Province, where about 5 species are distributed, including *F. pentaphylla*, *F. chinensis*, *F. nilgerrensis*, *F. corymbosa*, and *F.

gracilis. The third is northeast China, mainly including Jilin Province, Heilongjiang Province, Liaoning Province, and Inner Mongolia Autonomous Region, where 3 species are distributed, including *F. mandschurica*, *F. vesca*, and *F. orientalis*. The fourth is the Xinjiang Autonomous Region, where 2 species are distributed, including *F. viridis* and *F. vesca*.

中国原产有15个野生草莓种类，包括10个二倍体种和5个四倍体种。对于每个草莓种类而言，都有其自己的分布范围，有的很狭窄，有的则很广泛（表3-1）。例如，台湾草莓在全世界范围内也只分布在台湾省，绿色草莓在我国只分布在新疆，高原草莓只分布在高原藏区，峨眉草莓目前发现只分布在四川和贵州，裂萼草莓只分布在西藏和云南，森林草莓虽然在全世界分布规范，但在我国只分布在新疆、黑龙江、吉林等少数省份，并不广泛；而黄毛草莓则分布在四川、云南、贵州、陕西、重庆、西藏、湖北、湖南等很广泛的地区，中国草莓分布也较广泛，在青海、甘肃、四川、湖北、陕西、河南等西北、西南、中部地区均有分布。

There are 15 wild strawberry species native to China, including 10 diploid species and 5 tetraploid species. There is a distribution range for each species; some are narrow, but others are very wide (Table 3-1). For example, *F. hayatai* is only distributed in Taiwan Province, China; *F. viridis* is only distributed in Xinjiang Autonomous Region in China; *F. tibetica* is only distributed in Xizang and near sites; *F. emeiensis* is only distributed in Sichuan and Guizhou Provinces; *F. daltoniana* is only distributed in Xizang and Yunnan Province; although *F.*

表3-1 中国原产野生草莓种类的分布省区
Table 3-1 The different province distribution of *Fragaria* species native to China

倍性 Ploidy	种 Species	中国分布 Distribution in China
二倍体 Diploid 2n=2x=14	五叶草莓（*F. pentaphylla*）	四川、甘肃、陕西、青海 Sichuan, Gansu, Shaanxi, Qinghai
	森林草莓（*F. vesca*）	新疆、吉林、黑龙江 Xinjiang, Jilin, Heilongjiang
	西藏草莓（*F. nubicola*）	西藏 Xizang
	黄毛草莓（*F. nilgerrensis*）	云南、四川、重庆、陕西、贵州、湖南、湖北、西藏、广西 Yunnan, Sichuan, Chongqing, Shaanxi, Guizhou, Hunan, Hubei, Xizang, Guangxi
	台湾草莓（*F. hayatai*）	台湾 Taiwan
	裂萼草莓（*F. daltoniana*）	西藏、云南 Xizang, Yunnan
	绿色草莓（*F. viridis*）	新疆 Xinjiang
	东北草莓（*F. mandschurica*）	吉林、黑龙江、内蒙古、辽宁 Jilin, Heilongjiang, Inner Mongolia, Liaoning
	中国草莓（*F. chinensis*）	青海、甘肃、四川、湖北、陕西、河南 Qinghai, Gansu, Sichuan, Hubei, Shaanxi, Henan
	峨眉草莓（*F. emeiensis*）	四川、贵州 Sichuan, Guizhou
四倍体 Tetraploid 2n=4x=28	东方草莓（*F. orientalis*）	吉林、黑龙江、内蒙古 Jilin, Heilongjiang, Inner Mongolia
	西南草莓（*F. moupinensis*）	云南、贵州、四川 Yunnan, Guizhou, Sichuan
	伞房草莓（*F. corymbosa*）	山西、河北、甘肃、陕西、河南 Shanxi, Hebei, Gansu, Shaanxi, Henan
	纤细草莓（*F. gracilis*）	青海、宁夏、四川 Qinghai, Ningxia, Sichuan
	高原草莓（*F. tibetica*）	西藏 Xizang

vesca is widely distributed in many countries in the world, it is only distributed in Xinjiang, Heilongjiang, and Jilin Provinces in China. On the contrary, *F. nilgerrensisi* is widely distributed in Sichuan, Yunnan, Guizhou, Shaanxi, Chongqing, Xizang, Hubei, Hunan and other provinces in Southwest and Central China, but *F. chinensis* is widely distributed in Qinghai, Gansu, Sichuan, Hubei, Shaanxi, Henan and other provinces in Northwest, Southwest and Central China.

根据我们的调查，中国至少21个省区分布有野生草莓（表3-2）。其中，四川省分布种类最多，至少分布有7个种；其次是西藏和云南，分别至少分布有5个和4个种；吉林、黑龙江、陕西、甘肃、贵州、青海、新疆、河南、湖北分别至少分布有2~3个种；而河北、山西、内蒙古、辽宁、重庆、湖南、宁夏、台湾大约分布有1个种。但我国东南部沿海地区则几乎没有野生草莓分布。根据我们的调查和有关植物志记载，我国如下省区没有野生草莓分布：江苏、浙江、上海、安徽、山东、福建、海南、浙江、江西、北京、天津、香港、澳门。

表3-2 中国主要省份分布的野生草莓种类

Table 3-2 The *Fragaria* species distributed in main provinces of China

序号 No.	省区 Provinces	种 Species
1	四川 Sichuan	黄毛草莓、五叶草莓、中国草莓、峨眉草莓、纤细草莓、伞房草莓、西南草莓 *F. nilgerrensis, F. pentaphylla, F. chinensis, F. emeiensis, F. gracilis, F. corymbosa, F. moupinensis*
2	西藏 Xizang	西藏草莓、裂萼草莓、黄毛草莓、西南草莓、高原草莓 *F. nubicola, F. daltoniana, F. nilgerrensis, F. moupinensis, F. tibetica*
3	云南 Yunnan	黄毛草莓、裂萼草莓、中国草莓、西南草莓 *F. nilgerrensis, F. daltoniana, F. chinensis, F. moupinensis*
4	吉林 Jilin	东北草莓、森林草莓、东方草莓 *F. mandschurica, F. vesca, F. orientalis*
5	黑龙江 Heilongjiang	东北草莓、森林草莓、东方草莓 *F. mandschurica, F. vesca, F. orientalis*
6	陕西 Shaanxi	中国草莓、五叶草莓、黄毛草莓 *F. chinensis, F. pentaphylla, F. nilgerrensis*
7	甘肃 Gansu	中国草莓、五叶草莓、伞房草莓 *F. chinensis, F. pentaphylla, F. corymbosa*
8	贵州 Guizhou	峨眉草莓、西南草莓 *F. nilgerrensis, F. emeiensis, F. moupinensis*
9	青海 Qinghai	中国草莓、纤细草莓 *F. chinensis, F. gracilis*
10	新疆 Xinjiang	绿色草莓、森林草莓 *F. viridis, F. vesca*
11	河南 Henan	中国草莓、伞房草莓 *F. chinensis, F. corymbosa*
12	湖北 Hubei	黄毛草莓、中国草莓 *F. nilgerrensis, F. chinensis*
13	内蒙古 Inner Mongolia	东北草莓、东方草莓 *F. mandschurica, F. orientalis*
14	河北 Hebei	伞房草莓 *F. corymbosa*
15	山西 Shanxi	伞房草莓 *F. corymbosa*
16	宁夏 Ningxia	中国草莓 *F. chinensis*
17	辽宁 Liaoning	东北草莓 *F. mandschurica*
18	重庆 Chongqing	黄毛草莓 *F. nilgerrensis*
19	湖南 Hunan	黄毛草莓 *F. nilgerrensis*
20	广西 Guangxi	黄毛草莓 *F. nilgerrensis*
21	台湾 Taiwan	台湾草莓 *F. hayatai*

According to our investigation, wild strawberries are naturally distributed in at least 21 provinces or autonomous regions in China (Table 3–2). Among them, Sichuan Province has the most species (8 species). Next are Xizang and Yunnan Province, having 5 and 4 species, respectively. Two to 3 species are distributed in Qinghai, Jilin, Heilongjiang, Shaanxi, Gansu, Guizhou, Xinjiang, Henan and Hubei Provinces. Only 1 species is distributed in Hebei, Shanxi, Inner Mongolia, Liaoning, Chongqing, Hunan and Taiwan Provinces. However, there are almost no wild strawberries in the eastern and southern coastal areas in China. According to our investigation and the records of relative flora, there is no wild strawberry distribution in the following provinces or cities in China: Jiangsu, Zhejiang, Shanghai, Anhui, Shandong, Fujian, Hainan, Zhejiang, Jiangxi, Beijing, Tianjin, Hong Kong, and Macao.

3.2 中国野生草莓各种类的分布
Distribution of various wild strawberry species in China

3.2.1 五叶草莓 *F. pentaphylla* Lozinsk.

五叶草莓在中国西部、西南地区分布广泛，尤其在秦巴山区分布最多。主要分布于四川、甘肃、陕西、青海等省，如四川成都市朝天区、广元市、茂县、都江堰市、南江县等地，甘肃天水市、西和县、礼县、清水县、康县、武都县、麦积山等地，陕西宁强县、略阳县、汉中市、郑县等地。

五叶草莓中，白果类型分布最为广泛，粉果和红果类型在野外群体较小。

F. pentaphylla is widely distributed in western and southwest China, especially in the Qinba Mountain area. It is mainly distributed in Sichuan, Gansu, Shaanxi, and Qinghai Provinces. For example, Chaotian District in Chengdu City, Guangyuan City, Maoxian County, Dujiangyan City, and Nanjiang County in Sichuan Province; Tianshui City, Xihe County, Lixian County, Qingshui County, Kangxian County, Wudu County, and Maiji Mountain in Gansu Province; Ningqiang County, Lueyang County, Hanzhong City, and Zhengxian County in Shaanxi Province.

The wild population of white-fruited type of *F. pentaphylla* is most widely distributed, while the pink- and red-fruited types were relatively rare.

3.2.2 森林草莓 *F. vesca* L.

森林草莓在我国只分布于新疆、黑龙江、吉林等少数省份，并不广泛，如新疆乌鲁木齐，黑龙江尚志、伊春，吉林蛟河。

在黑龙江发现有红果类型、白果类型、四季型森林草莓。但目前在新疆、吉林则只发现分布有红果森林草莓。

F. vesca is only distributed in Xinjiang Autonomous Region, Heilongjiang Province, and Jilin Province in China. For example, Urumqi City in Xinjiang Autonomous Region, Shangzhi City and Yichun City in Heilongjiang Province, Jiaohe City in Jilin Province.

The red-fruited type and white-fruited type of *F. vesca* were found in Heilongjiang Province. The variety of *F. vesca* var. *semperflorens* is also distributed in Heilongjiang Province. Only the red-fruited type of *F. vesca* was distributed in Xinjiang Autonomous Region and Jilin Province.

3.2.3 西藏草莓 *F. nubicola* Lindl.

西藏草莓分布于我国西藏。我们调查发现在西藏日喀则亚东县、聂拉木县、林芝县、错那县、樟木县、隆子县、洛扎县分布有西藏草莓的红果类型，在西藏波密县、墨脱县、林芝县、察隅县则分布有西藏草莓的白果类型。

F. nubicola is distributed in Xizang Autonomous Region. The red-fruited type of *F. nubicola* was found in Yadong County, Nyalam County, Nyingchi County, Cuona County, Zhangmu County, Longzi County and Luozha County, while the white-fruited type of *F. nubicola* was found in Bomi County, Motuo County, Nyingchi County and Chayu County.

3.2.4 黄毛草莓 *F. nilgerrensis* Schlechtendal ex J. Gay

黄毛草莓广泛分布于我国西南部，如云南、贵州、四川、重庆、陕西、西藏、湖北、湖南等省，如云南丽江市、昆明市、大理市、迪庆州、会泽县、定武县、贡山县、玉龙雪山等地，贵州六盘水市、贵阳市、遵义市、长顺县、黎平县等地，四川叙永县、古蔺县、合江县、南江县、广安市、万源市、眉山市、凉山州、峨眉山等地，重庆金佛山等地，陕西宁强县等地，西藏林芝县等地，湖北神农架、利川市、巴东县、保康县、房县等地，广西隆林县、龙胜县等地；湖南省石门县等地。

F. nilgerrensis is widely distributed in Southwest China, including Yunnan, Guizhou, Sichuan, Chongqing, Shaanxi, Xizang, Hubei, and Hunan Provinces, such as Lijiang City, Kunming City, Dali City, Diqing Prefecture, Huize County, Dingwu County, Gongshan County, and Yulong Snow Mountain in Yunnan Province; Liupanshui City, Guiyang City, Zunyi City, Changshun County, and Liping County in Guizhou Province; Xuyong County, Gulin County, Hejiang County, Nanjiang County, Guang'an City, Wanyuan City, Meishan City, Liangshan Prefecture, and Emei Mountains in Sichuan Province; Jinfo Mountain in Chongqing City; Ningqiang County in Shaanxi Province; Nyingchi County in Xizang; Shennongjia, Lichuan City, Badong County, Baokang County, and Fang County in Hubei Province; Longlin County and Longsheng County in Guangxi Province; Shimen County in Hunna Province.

3.2.5 裂萼草莓 *F. daltoniana* Gay

裂萼草莓分布于西藏、云南高海拔山区，如西藏的聂拉木县、定结县、墨脱县等地，云南的独龙江州贡山县、怒江州泸水市等地。

F. daltoniana is distributed in the mountain area with the high altitude in Xizang Autonomous Region and Yunnan Province. It is distributed in Nyalam County, Dingjie County and Metuo County in Xizang Autonomous Region, and Gongshan County in Dulongjiang Prefecture, and Lushui City in Nujiang Prefecture in Yunnan Province.

3.2.6 绿色草莓 *F. viridis* Duch.

绿色草莓在中国仅仅分布于新疆，如新疆塔城市、伊犁市、乌鲁木齐市、昭苏县、特克斯县、天山等地。

F. viridis is only distributed in Xinjiang Autonomous Region in China, such as Tacheng City, Ili City, Urumqi City, Zhaosu County, Tekes County, and Tianshan Mountains.

3.2.7 东北草莓 *F. mandschurica* Staudt

东北草莓分布于中国东北地区，包括吉林、黑龙江、内蒙古、辽宁省。如吉林安图县、敦化市、长白山等地，黑龙江尚志市、克山县、漠河县、伊春市、黑河市、加格达奇市等地，内蒙呼伦贝尔盟鄂伦春旗、呼伦贝尔盟牙克石市、赤峰市阿鲁科尔沁旗等地，辽宁丹东市等地。

F. mandschurica is distributed in northeast China, including Jilin, Heilongjiang, Inner Mongolia, and Liaoning Provinces. Such as Antu County, Dunhua City, and Changbai Mountains in Jilin Province; Shangzhi City, Keshan County, Mohe County, Yichun City, Heihe City, Jiagdaqi City in Heilongjiang Province; Hulunbuir League, Yakeshi City of Olunchun Banner and Alukerqin Banner of Chifeng City in Inner Mongolia Autonomous Region; Dandong City in Liaoning Province.

3.2.8 中国草莓 *F. chinensis* Lozinsk.

中国草莓分布较广泛，在甘肃、青海、宁夏、四川、陕西、湖北、河南等西北、西南、中部地区均有分布，如甘肃天水市、岷县、西和县、礼县、清水县、合作市、临夏市、庆阳市、莲花山、麦积山等地，青海西宁市、互助县、湟中区等地；陕西陇县、宁陕县等地，四川阿坝州九寨沟等地，湖北神农架等地，宁夏六盘山市、银川市等地。

F. chinensis is widely distributed in northwest, southwest and central regions in China, such as Gansu, Qinghai, Ningxia, Sichuan, Hubei, Shaanxi, and Henan Provinces. Such as Tianshui City, Minxian County, Xihe County, Lixian County, Qingshui County, Hezuo City, Linxia City, Qingyang City, Lianhua Mountain, and Maiji Mountain in Gansu Province; Xining City, Huzhu County, and Huangzhong District of Xining City in Qinghai Province; Longxian County and Ningshan County in Shaanxi Province; Jiuzhaigou County of Aba Prefecture in Sichuan Province; Shennongjia Forest Region in Hubei Province; Liupanshan City and Yinchuan City in Ningxia Autonomous Region.

3.2.9 台湾草莓 *F. hayatai* Staudt

台湾草莓仅仅分布于中国台湾省，主要在台湾中部的玉山。

F. hayatai is only distributed in Taiwan Province of China, mainly in Yushan Mountains located in central Taiwan.

3.2.10 峨眉草莓 *F. emeiensis* Jia J. Lei

峨眉草莓是2021年发表的一个新种，分布于四川、贵州省，如四川乐山市、彭州市、康定县、泸定县、大邑县、乡城县、石棉县、峨眉山、西岭雪山、贡嘎山、贵州梵净山等地。

F. emeiensis, a new species published in 2021, is distributed in Sichuan Province and Guizhou Province. For example, Leshan City, Pengzhou City, Kangding County, Luding County, Dayi County, Xiangcheng County, Shimian County, Mount Emei, Xiling Snow Mountain, and Gongga Mountain in Sichuan Province; Mount Fanjing in Guizhou Province.

3.2.11 东方草莓 *F. orientalis* Lozinsk.

东方草莓分布于我国东北寒冷地区，包括黑龙江省、吉林省，如黑龙江尚志市、七台河市、牡丹江市、双鸭山市、密山市、林口县、林海县等地，吉林珲春市、白山县、长白山等地。但我们通过多年的调查一直未在辽宁省找到东方草莓。

F. orientalis is distributed in the cold areas of northeast China, including Heilongjiang Province and Jilin

Province, such as Shangzhi City, Qitaihe City, Mudanjiang City, Shuangyashan City, Mishan City, Linkou County, and Linhai County in Heilongjiang Province; Hunchun City, Baishan County, and Changbai Mountains in Jilin Province. But we have not found *F. orientalis* in Liaoning Province up to now.

3.2.12 西南草莓 *F. moupinensis* (Franch) Card.

西南草莓分布于我国西南地区，包括云南、贵州等省，如云南迪庆州、香格里拉市、丽江市、德钦县、中甸县、玉龙雪山等地，贵州六盘水市等地。

F. moupinensis is distributed in southwest China, including Yunnan Province and Guizhou Province. For example, Diqing Prefecture, Shangri-La City, Lijiang City, Deqin County, Zhongdian County, and Yulong Snow Mountain in Yunnan Province; Liupanshui City in Guizhou Province.

3.2.13 伞房草莓 *F. corymbosa* Lozinsk.

伞房草莓主要分布于山西、河北、甘肃、陕西、河南等省，如山西五台山、沁源县、阳高县、岚县等地，河北涞源县、石家庄市等地。

F. corymbosa is mainly distributed in Shaanxi, Hebei, Gansu, Shaanxi, Sichuan and Henan Provinces. For example, Wutai Mountain, Qinyuan County, Yanggao County and Lanxian County in Shanxi Province; Laiyuan County and Shijiazhuang City in Hebei Province.

3.2.14 纤细草莓 *F. gracilis* Lozinsk.

纤细草莓主要分布于青海、四川等省，如青海大通县、互助县等地，四川阿坝州松潘县等地。

F. gracilis is mainly distributed in Qinghai Province and Sichuan Province. For example, Datong County and Huzhu County in Qinghai Province; Songpan County of Aba Prefecture in Sichuan Province.

3.2.15 高原草莓 *F. tibetica* Staudt et Dickoré

高原草莓分布于西藏，如西藏林芝县、米林县、南迦巴瓦峰等地。云南和四川可能也有分布。

F. tibetica is distributed in Xizang Autonomous Region, such as Nyingchi County, Milin County, and Namcha Barwa NW slope. It might also distribute in Yunnan Province and Sichuan Province.

中国野生草莓的分类

Taxonomy of wild strawberry resources distributed in China

4.1 中国野生草莓的种类

The wild strawberry species distributed in China

全世界草莓属26个种中，25个为野生种、1个为栽培种。中国是世界上草莓野生种类资源最丰富的国家，自然分布有15个种，包括10个二倍体种和5个四倍体种。10个二倍体种是中国草莓、裂萼草莓、峨眉草莓、东北草莓、黄毛草莓、台湾草莓、西藏草莓、五叶草莓、森林草莓、绿色草莓，5个四倍体种是伞房草莓、纤细草莓、西南草莓、东方草莓、高原草莓。其中，峨眉草莓为新发现种（Qiao Qin et al, 2021）。此外，近年来还发现我国东北分布有自然五倍体野生草莓（Jiajun Lei et al, 2005）。全世界草莓属14个二倍体种中，有10个二倍体种分布在我国，占绝大多数，而且全世界草莓属所有5个四倍体种全部几乎只分布在我国，只有少数分布在周边邻国。

There are 26 recognized *Fragaria* species in the world, including 25 wild species and 1 cultivated species. China has more wild strawberry resources than any other country in the world. There are 15 wild species distributed in China, including 10 diploid species: *F. chinensis* Lozinsk., *F. daltoniana* Gay, *F. emeiensis* Jia J. Lei, *F. hayatai* Staudt, *F. mandschurica* Staudt, *F. nilgerrensis* Schlecht., *F. nubicola* Lindl., *F. pentaphylla* Lozinsk., *F. vesca* L., *F. viridis* Duch., and 5 tetraploid species: *F. corymbosa* Lozinsk., *F. gracilis* A. Los., *F. moupinensis* (Franch) Card., *F. orientalis* Lozinsk, and *F. tibetica* Staudt et Dickoré. Among them, *F. emeiensis* Jia J. Lei is a newly discovered species (Qiao Qin et al, 2021). Besides, a wild pentaploid strawberry genotype has been discovered in Northeast China in recent years (Jiajun Lei et al, 2005). Of all the 14 diploid species in the world, 10 are distributed in China, and all 5 tetraploid species in the world are almost only distributed in China, or only few distributed in neighboring countries.

4.2 中国野生草莓种类分类检索表

The dichotomous key of wild strawberry species in China

为了更好地进行中国原产草莓属植物15个种类的鉴定与分类，我们编制了其分种检索表，便于简单快速地进行分类。

In order to identify the 15 wild strawberry species native to China, we gave the dichotomous key to classify them easily and rapidly.

表4-1　中国及世界草莓属（*Fragaria*）植物的种类及其分布

Table 4-1 The genus *Fragaria* species and their distribution in China and in the world

倍性 Ploidy	种类 Species	世界分布 Distribution
二倍体 （2n=2x=14）	两季草莓（*F. × bifera* Duch.）	法国、德国 France, Germany
	布哈拉草莓（*F. bucharica* Losinsk.）	喜马拉雅山西部 Western Himalayas
	中国草莓（*F. chinensis* Losinsk.）*	中国西中部 West-central China
	裂萼草莓（*F. daltoniana* J. Gay）*	喜马拉雅山 Himalayas
	峨眉草莓（*F. emeiensis* Jia J. Lei）*	中国西南部 Southwest China
	台湾草莓（*F. hayatai* Staudt）*	中国台湾 Taiwan, China
	饭沼草莓（*F. iinumae* Makino）	日本 Japan
	东北草莓（*F. mandschurica* Staudt）*	中国东北 Northeast China
	黄毛草莓（*F. nilgerrensis* Schlecht.）*	亚洲西南部 Southeastern Asia
	日本草莓（*F. nipponica* Makino）	日本 Japan
	西藏草莓（*F. nubicola* Lindl.）*	喜马拉雅山 Himalayas
	五叶草莓（*F. pentaphylla* Losinsk.）*	中国西南、西北部 Southwest and Northwest China
	森林草莓（*F. vesca* L.）*	欧洲、亚洲西部、北美 Europe, Asia west, North America
	绿色草莓（*F. viridis* Duch.）*	欧洲、东亚 Europe and East Asia
四倍体 （2n=4x=28）	伞房草莓（*F. corymbosa* Losinsk.）*	中国西中部 West-central China
	纤细草莓（*F. gracilis* A. Los.）*	中国西北部 Northwest China
	西南草莓（*F. moupinensis* (Franch.) Card.）*	中国西南部 Southwest China
	东方草莓（*F. orientalis* Losinsk.）*	中国、俄罗斯远东 China and Russian Far East
	高原草莓（*F. tibetica* Staudt et Dickoré）*	中国西南部 Southwest China
五倍体 （2n=5x=35）	布氏草莓（*F. bringhurstii* Staudt）	美国加州 California, USA
六倍体 （2n=6x=42）	麝香草莓（*F. moschata* Duch.）	欧洲–西伯利亚 Euro-Siberia
八倍体 （2n=8x=56）	凤梨草莓（*F. ananassa* Duch.）	世界各国均引种栽培 Cultivated worldwide
	智利草莓（*F. chiloensis* (L.) Duch.）	南美（智利）、北美西部 South America (Chile), Western N. America
	弗州草莓（*F. virginiana* Duch.）	北美 North America
十倍体 （2n=10x=70）	喀斯喀特草莓（*F. cascadensis* Hummer）	美国俄勒冈州 Oregon, USA
	择捉草莓（*F. iturupensis* Staudt）	太平洋西北部的择捉岛 Iturup Island in Northwest Pacific

注：符号*表示中国自然分布有该种。Note: The symbol * indicates that this species naturally distributes in China.

表4-2　中国原产野生草莓分种检索

Table 4-2 The dichotomous key of *Fragaria* species native to China

1 匍匐茎合轴 Runner sympodial
2 花两性 Plant hermaphroditic (2n=2x=14)
3 叶柄，匍匐茎和花梗被开展绒毛Petiole, runner and peduncle covered spreading hair
4 果实白色或红色，种子凹，萼片紧贴Fruit white or red; achene extremely sunken; calyx clasping
5 果实白色，花瓣白色，匍匐茎绿色或浅红色⋯⋯⋯⋯⋯⋯⋯⋯⋯⋯⋯⋯1 黄毛草莓 *F. nilgerrensis*   Fruit white; petal white; runner green or light red
5 果实红色，花瓣基部明显红色，匍匐茎深红色⋯⋯⋯⋯⋯⋯⋯⋯⋯　2 台湾草莓 *F. hayatai*   Fruit red; petal white but red at base; runner deep red
4 果实红色，种子凸，萼片平展或稍反折 ⋯⋯⋯⋯⋯⋯⋯⋯⋯⋯ 3 东北草莓 *F. mandschurica*   Fruit red; achene raised; calyx spreading or lightly reflexed
3 叶柄被开展绒毛，匍匐茎和花梗被紧贴绒毛   Petiole covered spreading hair, runner and peduncle covered appressed hair
6 叶片绿色，非革质，无光泽，果实圆锥形，副萼全缘⋯⋯⋯⋯⋯⋯⋯⋯　4 森林草莓 *F. vesca*   Leaf green, non-coriaceous, non-shiny; fruit conic; calycle with entire apex
6 叶片深绿色，革质，有光泽，果实长纺锤形，副萼2或3裂⋯⋯⋯⋯⋯　5 裂萼草莓 *F. daltoniana*   Leaf dark green, coriaceous, shiny; fruit spindly; calycle with 2 or 3 lobed apexes
2 花单性，极稀两性Plant dioecious, rarely sub-dioecious (2n=4x=28)⋯⋯⋯⋯⋯6 东方草莓 *F. orientalis*
1 匍匐茎单轴Runner monopodial
7 花两性Plant hermaphroditic (2n=2x=14)
8 三小叶，稀五小叶Leaf trifoliate, rarely pinnately quinquefoliolate
9 三小叶，果实浅绿色或带红晕，硬，萼片紧贴，种子凸⋯⋯⋯⋯⋯⋯7 绿色草莓 *F. viridis*   Leaf trifoliate, fruit light green to reddish-green, hard; calyx clasping; achene raised
9 三小叶，稀五小叶，果实红色，软，萼片开展，种子凹⋯⋯⋯⋯⋯8 中国草莓 *F. chinensis*   Leaf trifoliate, rarely pinnately quinquefoliolate; fruit red, soft; calyx spreading; seed sunken
8 五小叶，稀三小叶 Leaf pinnately quinquefoliolate, rarely trifoliate
10 叶柄和匍匐茎被开展绒毛，花梗被稀疏绒毛，萼片反折⋯⋯⋯⋯ 9 五叶草莓 *F. pentaphylla*   Petiole and runner covered spreading hair, but sparse hairs on peduncle; calyx reflexed
10 叶柄、匍匐茎和花梗被紧贴绒毛，萼片紧贴或开展   Petiole, runner and peduncle covered appressed hair; calyx clasping or spreading
11 叶柄、匍匐茎较细，叶片较薄，叶片浅绿色，萼片紧贴 ⋯⋯⋯⋯ 10 西藏草莓 *F. nubicola*   Petiole, runner, leaf relatively thin, light green; calyx clasping
11 叶柄、匍匐茎较粗，叶片较厚，叶片深绿色，萼片开展⋯⋯⋯⋯ 11 峨眉草莓 *F. emeiensis*   Petiole, runner, leaf relatively thick, deep green; calyx spreading
7 花单性，极稀两性Plant dioecious, rarely sub-dioecious (2n=4x=28)
12 三小叶，或三小叶极稀五小叶Leaf trifoliate, or leaf trifoliate and rarely quinquefoliolate
13 三小叶，植株小、细弱，果实很小Leaf trifoliate; plant small, thin; fruit very small ⋯⋯⋯12 纤细草莓 *F. gracilis*
13 三小叶极稀五小叶，植株相对高，果实相对大Leaf trifoliate, rarely quinquefoliolate; plant relatively high; fruit relatively large⋯⋯⋯⋯⋯⋯⋯⋯⋯⋯⋯⋯⋯⋯⋯⋯⋯13 伞房草莓 *F. corymbosa*
12 五小叶，稀三小叶Leaf pinnately quinquefoliolate, rarely trifoliate
14 叶柄、匍匐茎和花梗被开展绒毛Petiole, runner and peduncle covered spreading hair   ⋯⋯⋯⋯⋯⋯⋯⋯⋯⋯⋯⋯⋯⋯⋯⋯⋯⋯⋯⋯⋯⋯⋯⋯⋯⋯⋯14 西南草莓 *F. moupinensis*
14 叶柄、匍匐茎和花梗被紧贴或上斜绒毛Petiole, runner and peduncle covered appressed or ascending hair   ⋯⋯⋯⋯⋯⋯⋯⋯⋯⋯⋯⋯⋯⋯⋯⋯⋯⋯⋯⋯⋯⋯⋯⋯⋯⋯⋯15 高原草莓 *F. tibetica*

4.3 中国野生草莓种类按倍性分类

The taxonomy of wild strawberry species in China according to ploidy

草莓属植物倍性丰富，是世界上被子植物中倍性最丰富的属之一，其自然分布的种类分别有二倍体、四倍体、五倍体、六倍体、八倍体、十倍体6种倍性。中国原产野生草莓15个种类中有2种倍性，包括10个二倍体种（$2n=2x=14$）、5个四倍体种（$2n=4x=28$）。我国东北还自然分布有五倍体野生草莓。

Fragaria is one of the genera with the most ploidies in the plant kingdom in the world, and there are natural diploid, tetraploid, pentaploid, hexaploid, octoploid, and decaploid species. Of the 15 native wild strawberry species in China, 10 diploid species ($2n=2x=14$) and 5 tetraploid species ($2n=4x=28$) were recorded. A wild pentaploid strawberry genotype has been also discovered in Northeast China in recent years.

（1）中国分布的10个二倍体种（$2n=2x=14$）：中国草莓、裂萼草莓、峨眉草莓、东北草莓、黄毛草莓、台湾草莓、西藏草莓、五叶草莓、森林草莓、绿色草莓。

The 10 diploid species distributed in China（$2n=2x=14$）：*F. chinensis*, *F. daltoniana*, *F. emeiensis*, *F. hayata*, *F. mandschurica*, *F. nilgerrensisi*, *F. nubicola*, *F. pentaphylla*, *F. vesca*, and *F. viridis*.

（2）中国分布的5个四倍体种（$2n=4x=28$）：伞房草莓、纤细草莓、西南草莓、东方草莓、高原草莓。

The 5 tetraploid species distributed in China ($2n=4x=28$)：*F. corymbosa*, *F. gracilis*, *F. moupinensism*, *F. orientalis*, and *F. tibetica*.

4.4 中国野生草莓种类按匍匐茎分支类型分类

The taxonomy of wild strawberry species in China according to runner branching

按照其结构发生类型，草莓属植物的匍匐茎可以分为合轴和单轴两大类。草莓匍匐茎合轴和匍匐茎单轴可以通过如下情形来辨认：匍匐茎上仅偶数节着生苗者为合轴，匍匐茎上奇数节和偶数节均着生苗者（第一节除外）为单轴。

According to the structure and development of runner in the genus *Fragaria*, it can be divided into two branching types: sympodial and monopodial. The sympodial runner and monopodial runner can be recognized according to producing plantlets styles on even or odd nodes on runner in strawberry. Sympodial: plantlets produced only from even nodes on the runner; Monopodial: plantlets produced from both odd and even nodes on the runner (except for the first node).

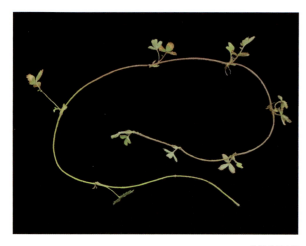 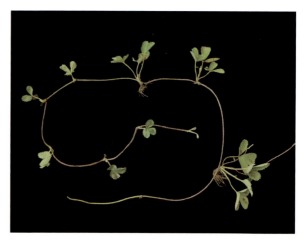

单轴匍匐茎 Monopodial runner

合轴匍匐茎 Sympodial runner

中国原产野生草莓的两种匍匐茎分支类型
Two runner branching types of *Fragaria* spp. native to China.

中国原产的15个种中，6个种为合轴匍匐茎，9个种为单轴匍匐茎。

Among 15 *Fragaria* species native to China according to the runner branching types, 6 species are sympodial and 9 species are monopodial.

（1）中国分布的匍匐茎为合轴的6个种类：森林草莓、黄毛草莓、台湾草莓、裂萼草莓、东北草莓、东方草莓。这6个合轴匍匐茎种中，5个为二倍体（森林草莓、黄毛草莓、台湾草莓、裂萼草莓、东北草莓）、1个为四倍体（东方草莓）。

The 6 sympodial-runnered species distributed in China: *F. vesca, F. nilgerrensis, F. hayatai, F. daltoniana, F. mandschurica,* and *F. orientalis.* Of the 6 sympodial species, 5 are diploid (*F. vesca, F. nilgerrensis, F. hayatai, F. daltoniana,* and *F. mandschurica*) and 1 is tetraploid (*F. orientalis*).

（2）中国分布的匍匐茎为单轴的9个种类：中国草莓、峨眉草莓、西藏草莓、五叶草莓、绿色草莓、伞房草莓、纤细草莓、西南草莓、高原草莓。这9个单轴匍匐茎种中，5个为二倍体（中国草莓、峨眉草莓、西藏草莓、五叶草莓、绿色草莓）、4个为四倍体（西南草莓、伞房草莓、纤细草莓、高原草莓）。

The 9 monopodial-runnered species distributed in China: *F. chinensis*, *F. emeiensis*, *F. nubicola*, *F. pentaphylla*, *F. viridis*, *F. corymbosa*, *F. gracilis*, *F. moupinensis*, and *F. tibetica*. Of the 9 monopodial species, 5 are diploid (*F. chinensis*, *F. emeiensis*, *F. nubicola*, *F. pentaphylla*, and *F. viridis*), and 4 are tetraploid (*F. corymbosa*, *F. gracilis*, *F. moupinensis*, and *F. tibetica*).

4.5 中国野生草莓种类按小叶数分类

The taxonomy of wild strawberry species in China according to leaflet number

草莓属植物的叶片为复叶，分为三小叶和五小叶两种。我们平时常见的栽培草莓品种为凤梨草莓种，都是三小叶的，但在野生草莓种类中五小叶较为常见。有趣的是，在欧洲、南美洲、北美洲及亚洲的日本，其原产的野生草莓种类都是三小叶的，只有中国及个别周边邻国（如印度）原产有五小叶的野生草莓种类。

The leaf of strawberry is compound, which is divided into two types according to leaflet number: trifoliate leaf and quinquefoliolate leaf. All the strawberry cultivars belonging to *F. ananassa* are trifoliate. However, the quinquefoliolate leaf is common among wild strawberry species native to China. It is interesting that, in Europe, South America, North America, and even Japan in Asia, all the native wild strawberry species are trifoliate, but there are several quinquefoliolate-leafed wild species only in China and few neighboring countries (such as India).

中国分布的15个原生种中，8个种为三小叶、7个种为五小叶。

Among the 15 species native to China, 8 species are trifoliate and 7 species are quinquefoliolate.

（1）中国分布的叶片为三小叶的8个种类：森林草莓、黄毛草莓、台湾草莓、裂萼草莓、东北草莓、绿色草莓、东方草莓、纤细草莓。这8个三小叶中，6个为二倍体（森林草莓、黄毛草莓、台湾草莓、裂萼草莓、东北草莓、绿色草莓）、2个为四倍体（东方草莓、纤细草莓）。

The 8 trifoliate species distributed in China: *F. vesca*, *F. nilgerrensis*, *F. hayatai*, *F. daltoniana*, *F. mandschurica*, *F. viridis*, *F. orientalis*, and *F. gracilis*. Of the 8 trifoliate species, 6 are diploid (*F. vesca*, *F. nilgerrensis*, *F. hayatai*, *F. daltoniana*, *F. mandschurica*, and *F. viridis*) and 2 are tetraploid (*F. orientalis* and *F. gracilis*).

（2）中国分布的叶片为五小叶的7个种类：中国草莓、西藏草莓、五叶草莓、峨眉草莓、伞房草莓、西南草莓、高原草莓。这7个五小叶种中，4个为二倍体（中国草莓、西藏草莓、五叶草莓、峨眉草莓）、3个为四倍体（伞房草莓、西南草莓、高原草莓）。

The 7 quinquefoliolate species distributed in China: *F. chinensis*, *F. nubicola*, *F. pentaphylla*, *F. emeiensis*, *F. corymbosa*, *F. moupinensis*, and *F. tibetica*. Of the 7 quinquefoliolate species, 4 are diploid (*F. chinensis*, *F. nubicola*, *F. pentaphylla*, and *F. emeiensis*) and 3 are tetraploid (*F. corymbosa*, *F. moupinensis*, and *F. tibetica*).

值得注意的是，按小叶数的种类分类不如按匍匐茎进行分类那么准确，因为有些为五小叶种类的叶片也经常出现三小叶，即使名称叫五叶草莓的这个种中有时也有三小叶存在。这种情况下，我们通常将之描述成常五小叶、稀三小叶。此外，伞房草莓、西藏草莓和中国草莓则较难界定其小叶的数目，常常三小叶和五小叶并存。而匍匐茎的合轴和单轴性状很稳定，且易于辨认。

叶片为三小叶（左：东北草莓；右：绿色草莓）
Trifoliate leaf (Left: *F. mandschurica*; Right: *F. viridis*)

叶片为五小叶（左：五叶草莓；右：高原草莓）
Quinquefoliolate leaf (Left: *F. pentaphylla*; Right: *F. tibetica*)

中国原产野生草莓种的两种小叶数叶片类型
Two types of leaflet numbers of *Fragaria* spp. native to China

It is worth noting that classifying by the leaflet number is less accurate than by runner type, because some quinquefoliolate species often have trifoliate leaf at the same time, and even for the typical quinquefoliolate species, *F. pentaphylla*. In this case, 'often quinquefoliolate, rarely trifoliate' leaf is the description for this species. But in other case, it is more difficult to define the leaflet numbers for *F. corymbosa*, *F. nubicola* and *F. chinensis*, because both quinquefoliolate and trifoliate leaves emerge at the same time even if on the same plant. But the sympodial runner and monopodial runner is more stable and easier to identify than the leaflet number.

4.6 中国野生草莓种类按花性别分类
The taxonomy of wild strawberry species in China according to flower sex

中国原产的野生草莓15个种类中，10个二倍体种的花均为两性花，植株为雌雄同株；5个四倍体种的花均为单性花，植株为雌雄异株。观察还发现，在西南草莓、东方草莓中还存在亚雌雄异株。

Of the 15 *Fragaria* species native to China, the flowers of all 10 diploid species are bisexual and the plants are hermaphroditic. The flowers of all 5 tetraploid species are unisexual and the plants are dioecious. But there are also unisexual and bisexual flowers observed at the same time in the populations of *F. moupinensis*, and *F. orientalis*, which are sub-dioecious.

（1）中国分布的雌雄同株的10个种类：中国草莓、裂萼草莓、峨眉草莓、东北草莓、黄毛草莓、台湾草莓、西藏草莓、五叶草莓、森林草莓、绿色草莓。

The 10 hermaphroditic species distributed in China: *F. chinensis, F. daltoniana, F. emeiensis, F. hayatai, F. mandschurica, F. nilgerrensis, F. nubicola, F. pentaphylla, F. vesca,* and *F. viridis*.

（2）中国分布的雌雄异株的5个种类：伞房草莓、纤细草莓、西南草莓、东方草莓、高原草莓。

The 5 dioecious species distributed in China: *F. corymbosa, F. gracilis, F. moupinensis, F. orientalis,* and *F. tibetica*.

两性花（左：五叶草莓；右：西藏草莓）
Bisexual flowers (Left: *F. pentaphylla;* Right: *F. nubicola*)

单性花（高原草莓，左：雄花；右：雌花）
Unisexual flowers (*F. tibetica.* Left: male; Right: female)

中国原产野生草莓二倍体种和四倍体种的花性别
The flower sex of diploid and tetraploid *Fragaria* species native to China

表4-3　中国原产野生草莓种类的匍匐茎类型、小叶数及花性别

Table 4-3 The runner type, leaflet number and flower sex of *Fragaria* spp. native to China

倍性 Ploidy	种 Species	匍匐茎类型 Runner type	小叶数 Leaflet number	性别 Sex
二倍体 Diploid 2n=2x=14	五叶草莓（*F. pentaphylla*）	单轴 Monopodial	五小叶，偶有三小叶 Quinquefoliolate, rarely trifoliate	雌雄同株 Hermaphroditic
	森林草莓（*F. vesca*）	合轴 Sympodial	三小叶 Trifoliate	雌雄同株 Hermaphroditic
	西藏草莓（*F. nubicola*）	单轴 Monopodial	三小叶和五小叶 Trifoliate and quinquefoliolate	雌雄同株 Hermaphroditic
	黄毛草莓（*F. nilgerrensis*）	合轴 Sympodial	三小叶 Trifoliate	雌雄同株 Hermaphroditic
	台湾草莓（*F. hayatai*）	合轴 Sympodial	三小叶 Trifoliate	雌雄同株 Hermaphroditic
	裂萼草莓（*F. daltoniana*）	合轴 Sympodial	三小叶 Trifoliate	雌雄同株 Hermaphroditic
	绿色草莓（*F. viridis*）	单轴 Monopodial	三小叶 Trifoliate	雌雄同株 Hermaphroditic
	东北草莓（*F. mandschurica*）	合轴 Sympodial	三小叶 Trifoliate	雌雄同株 Hermaphroditic
	中国草莓（*F. chinensis*）	单轴 Monopodial	三小叶和五小叶 Trifoliate and quinquefoliolate	雌雄同株 Hermaphroditic
	峨眉草莓（*F. emeiensis*）	单轴 Monopodial	五小叶 Quinquefoliolate	雌雄同株 Hermaphroditic
四倍体 Tetraploid 2n=4x=28	伞房草莓（*F. corymbosa*）	单轴 Monopodial	三小叶和五小叶 Trifoliate and quinquefoliolate	雌雄异株，偶有雌雄同株 Hermaphroditic Dioecious, rarely sub-dioecious
	纤细草莓（*F. gracilis*）	单轴 Monopodial	三小叶 Trifoliate	雌雄异株 Dioecious
	西南草莓（*F. moupinensis*）	单轴 Monopodial	五小叶，偶有三小叶 Quinquefoliolate, rarely trifoliate	雌雄异株，偶有雌雄同株 Hermaphroditic Dioecious, rarely sub-dioecious
	东方草莓（*F. orientalis*）	合轴 Sympodial	三小叶 Trifoliate	雌雄异株，偶有雌雄同株 Hermaphroditic Dioecious, rarely sub-dioecious
	高原草莓（*F. tibetica*）	单轴 Monopodial	五小叶，偶有三小叶 Quinquefoliolate, rarely trifoliate	雌雄异株 Dioecious

第五章
Chapter 5

中国野生草莓种类的特点
Characteristics of wild strawberry species in China

中国是世界上野生草莓种类最多的国家。最近研究表明，中国自然分布有15个野生种，种类占世界草莓属植物26种的一半以上（雷家军等，2006；Lei et al，2014；Lei et al，2017；Qiao et al，2021）。这15个野生种包括10个二倍体种：中国草莓、裂萼草莓、峨眉草莓、东北草莓、黄毛草莓、台湾草莓、西藏草莓、五叶草莓、森林草莓、绿色草莓和5个四倍体种：伞房草莓、纤细草莓、西南草莓、东方草莓、高原草莓。现将我国原产的15个野生种及其变种性状简介如下。

中国草莓 F. chinensis　裂萼草莓 F. daltoniana　峨眉草莓 F. emeiensis　东北草莓 F. mandschurica

黄毛草莓 F. nilgerrensisi　绿色草莓 F. viridis　五叶草莓红果 Red-fruited F. pentaphylla　五叶草莓白果 White-fruited F. pentaphylla

西藏草莓红果 Red-fruited F. nubicola　西藏草莓白果 White-fruited F. nubicola　森林草莓红果 Red-fruited F. vesca　森林草莓白果 White-fruited F. vesca

伞房草莓 F. corymbosa　纤细草莓 F. gracilis　西南草莓 F. moupinensism　东方草莓 F. orientalis　高原草莓 F. tibetica

中国原产15个野生草莓种类及变种的花
The flowers of 15 wild strawberry species and their varieties native to China

China has the most abundant wild strawberry resources in the world. Recent studies have shown that there are 15 wild species naturally distributed in China, accounting for more than half of the genus *Fragaria* species in the world (Lei Jiajun et al., 2006; Lei Jiajun et al., 2013; Lei Jiajun et al., 2017; Qiao Qin et al., 2019). The 15 wild species include 10 diploid species: *F. chinensis* Losinsk., *F. daltoniana* J. Gay, *F. hayatai* Staudt, *F. mandschurica* Staudt, *F. nilgerrensis* Schlecht., *F. nubicola* Lindl., *F. pentaphylla* Losinsk., *F. vesca* L., *F. viridis* Duch., *F. emeiensis* Jia J. Lei, and 5 tetraploid species: *F. corymbosa* Losinsk., *F. gracilis* A. Los., *F. moupinensis* (Franch.) Card., *F. orientalis* Losinsk., and *F. tibetica* Staudt et Dickoré. The characteristics of 15 wild species and their varieties native to China are described as below.

中国草莓 *F. chinensis*　　裂萼草莓 *F. daltoniana*　　峨眉草莓 *F. emeiensis*　　东北草莓 *F. mandschurica*

黄毛草莓 *F. nilgerrensis*　　绿色草莓 *F. viridis*　　五叶草莓红果 Red-fruited *F. pentaphylla*　　五叶草莓白果 White-fruited *F. pentaphylla*

西藏草莓红果 Red-fruited *F. nubicola*　　西藏草莓白果 White-fruited *F. nubicola*　　森林草莓红果 Red-fruited *F. vesca*　　森林草莓白果 White-fruited *F. vesca*

伞房草莓 *F. corymbosa*　　纤细草莓 *F. gracilis*　　西南草莓 *F. moupinensism*　　东方草莓 *F. orientalis*　　高原草莓 *F. tibetica*

中国原产15个野生草莓种类及变种的果实
The fruits of 15 wild strawberry species and their varieties native to China

5.1 五叶草莓

F. pentaphylla Lozinsk.

植株高5~15 cm。羽状五小叶，下部的2片叶要远比上面的三出叶小，稀羽状三小叶，叶柄上常有2~3个耳叶。中心小叶具短柄，两边小叶无柄。叶片质地厚，椭圆形、长椭圆形或倒卵圆形，叶背面常呈现紫红色。叶柄、匍匐茎上被开展绒毛，但花梗上绒毛稀少。匍匐茎较粗，单轴分枝。花序高于叶面，每花序上花朵数少，常2~3朵。花瓣边缘常具波状，雄蕊不等长。果实卵球形，白色，亦有红色或粉红色，自然界中白果类型居多，白果香、略酸，红果或粉果无香味、酸。种子凹陷。宿萼反折。我国原产五叶草莓的果实有白果、粉果和红果3种类型。二倍体2n=2x=14。

The plant is 5~15 cm high. The leaf is pinnately quinquefoliolate (but the lower two leaves are much smaller than the upper three leaves), rarely trifoliate, often 2~3 auriculate leaves on petioles. The central leaflet is elliptic with long petiole, nearly glabrous adaxially and often purplish red abaxially. The central leaflet has a short petiolate, but lateral leaflets are sessile. The petioles and runners are covered with spreading hairs, but sparse hairs on peduncles. The runner is monopodial, thick. The inflorescence is cymose with 2 to 3 flowers and above the leaf surface. The petal is wedge at the apex and wavy at the edge. The stamens are unequal in length. The fruits are ovoid. The fruits are white, pink or red. The white fruit is aromatic and slightly acid, but the pink or red fruits are not aromatic and very acid. Achene is extremely sunken on the fruit surface. The calyx is reflexed. There are three fruit types (white, pink and red) in *F. pentaphylla* in China. Diploid (2n =2x= 14).

叶片背面常呈紫红色
Leaflet is often purplish red abaxially.

白果 White fruit

粉果 Pink fruit

红果 Red fruit

五叶草莓的植株、叶片、花和果实
Plants, leaves, flowers and fruits of *F. pentaphylla*

5.2 森林草莓

F. vesca L.

植株直立，高10～30 cm。羽状三小叶。楔状卵圆形、菱状卵圆形，边缘锯齿大而尖。小叶近无柄或仅中心小叶具短柄。叶柄上具开展绒毛。匍匐茎合轴分枝，具紧贴绒毛。花序上有花1～6朵。花序上从第一个分歧处往上绒毛均为紧贴，往下则直立或斜生。花瓣前端常具2～4个缺刻。雄蕊近等长，低于或平于雌蕊。果实长圆锥形、圆锥形、卵球形，红色或白色，肉软，汁液少，香味浓。种子凸于果面，红色（红果者）或黄色（白果者）。宿萼开展或微反折。中国原产的森林草莓有红果和白果类型。另外，黑龙江省还有四季结果型森林草莓。二倍体2n=2x=14。

The plant is erect with a 10~30 cm height. The leaf is trifoliate, cuneate-ovate, rhomboid-ovate with large and pointed sawtooth on the edge. The leaflet is almost sessile, or with short petiole only on the central leaflet. The petiole is covered with spreading hairs. The runner is sympodial branching and covered with appressed hairs except the first node. The inflorescence is often 1 to 6 flowers. The peduncle is covered with appressed hairs except from the first fork below. The petal often has 2~4 serrations in the apex. The filament is short and almost equal in length. The fruit is red or white, long conic, conic, or ovoid, white flesh with a pale yellow tint, highly aromatic, soft, hollow core, little juice. The achene is red (on the red fruit) or yellow (on the white fruit), and raised. The calyx is spreading or lightly reflexed. There are two fruit types (white and red) in *F. vesca* in China. The variety *F. vesca* var. *semperflorens* is also distributed in Heilongjiang Province. Diploid ($2n=2x=14$).

红果 Red fruit　　　　　　　　　　　　　白果 White fruit

森林草莓的植株、叶片、花和果实
Plants, leaves, flowers and fruits of *F. vesca*

5.3 西藏草莓
F. nubicola Lindl.

植株高4～20 cm。羽状五小叶，稀三小叶。小叶椭圆形或倒卵圆形，叶缘具尖锯齿，小叶无柄或具短柄。叶正面绿色，背面淡绿色。叶柄上绒毛紧贴，少数开展。匍匐茎较细，单轴分枝，其上绒毛紧贴。花序上花朵少，常1～4朵，花梗上绒毛紧贴。果实卵球形。种子凹。宿萼紧贴。中国原产的西藏草莓有红果和白果两种类型。二倍体2*n*=2*x*=14。

The plant is 4~20 cm high. The leaf is pinnately quinquefoliolate, rarely trifoliate, elliptic or obovate with very sharp teeth on the edge. The leaflet is almost sessile, or with short petiole. The petiole, runner and peduncle are covered with appressed hairs. The runner is filiform, and monopodial. The fruit is globose. The achene is sunken. The calyx is appressed. There are two fruit types (white and red) in *F. nubicola* in China. Diploid (2*n*=2*x*=14).

西藏草莓的植株、叶片、花和果实
Plants, leaves, flowers and fruits of *F. nubicola*

5.4 黄毛草莓

F. nilgerrensis Schlechtendal ex J. Gay

植株健壮，株高10～25 cm，羽状三小叶。叶片深绿色，厚，小叶倒卵圆形至圆形，前端平楔形。小叶近无柄或仅中心小叶具短柄。匍匐茎合轴分枝。匍匐茎、叶柄和花序梗上均被长而直立的棕黄色或白色绒毛，因此称为"黄毛草莓"。花瓣离生，雄蕊花丝较短。果半球形或扁球形，白色略黄或略带浅粉红色，常无味或味淡，但有的香味很浓。种子很小且数量多，着生密集，凹于果面。宿萼紧贴。黄毛草莓在野生草莓中开花几乎是最晚的。二倍体2n=2x=14。

The plant is robust with a 10~25 cm height. The leaf is trifoliate, deep green, thick, obovate or subround with cuneate apex. The leaflet is almost sessile, or with short petiole only on the central leaflet. The runner is sympodial. The petiole, runner and peduncle are covered with densely spreading hairs. The petal is obviously apopetalous. The fruit is globose, white with a pale yellow tint, slightly to strongly aromatic. The achene is very small and extremely sunken. The calyx is clasping. It blooms almost latest among all wild strawberry species. Diploid ($2n=2x=14$).

黄毛草莓的植株、叶片、花和果实
Plants, leaves, flowers and fruits of *F. nilgerrensis*

5.5 裂萼草莓

F. daltoniana Gay

植株细弱矮小，高4～6 cm。羽状三小叶，叶深绿色，革质，具光泽。中心小叶具小叶柄。匍匐茎纤细，合轴分枝。叶柄、匍匐茎、花梗具紧贴绒毛。每花序长1朵花。副萼片宽披针形，顶端2~3浅裂，因此称"裂萼草莓"。果相对稍大，呈纺锤形或长卵圆形，鲜红色，果肉海绵质，几乎无味。种子凹陷。宿萼开展。我国原产的裂萼草莓的果实有红果、白果、粉果3种类型。二倍体2n=2x=14。

The plant is slender with a 4~6 cm height. The leaf is trifoliate, shiny, coriaceous, with petiolate central leaflets. The petiole, runner and peduncle are covered with appressed hairs. The runner is filiform, sympodial branching. Often one flower per inflorescence. The calycle is wide lanceolate with two or three lobed apexes. The fruit is relatively big, spindly or cylindrical with spongy tasteless flesh. The calyx is spreading. There are three fruit types (red, white and pink) in *F. daltoniana* in China. Diploid (2n=2x=14).

裂萼草莓的植株、叶片、花和果实
Plants, leaves, flowers and fruits of *F. daltoniana*

5.6 绿色草莓

F. viridis Duch.

　　植株纤细，株高10～25 cm。羽状三小叶，叶深绿色，薄，椭圆形至卵圆形。叶柄上具开展绒毛。匍匐茎单轴分枝。花序直立，常4～10朵花明显高于叶面。归圃栽植时，群体花量大，单株花朵数多。花明显比森林草莓大，花丝细长，高于雌蕊，花丝比森林草莓更长、花药更大。花瓣在开花初期有时呈淡黄绿色，后期变为白色。果实成熟时呈淡绿色，阳面略红，扁圆形，硬，有清香味。种子大，平于果面。萼片相对大而长，宿萼紧贴，除萼难。通常在秋季能再次零星开花。二倍体2*n*=2*x*=14。

　　The plant is slender with a 10~25 cm height. The leaf is trifoliate, deep green, thin, elliptic to oval. The petiole is covered with thick spreading hairs. The runner is monopodial with appressed hairs except the first node. The inflorescence is erect and obviously taller than leaf level, often 4~10 flowers per inflorescence. The peduncle is covered with appressed hairs except the first fork below. The flower is significantly larger than *F. vesca*. The filament is slender and higher than the pistil, and its filament is longer and the anther is larger than *F. vesca*. The petal is sometimes yellowish green at the beginning of flowering, and then becomes white later. The fruit is light reddish green, oblate, and hard. The flesh is white with a pale yellowish green tint, and slightly aromatic. The achene is large, yellowish green, raised. The calyx is clasping and hard to separate. It often blooms again in autumn. Diploid (2*n*=2*x*=14).

绿色草莓的植株、叶片、花和果实
Plants, leaves, flowers and fruits of *F. viridis*

5.7 东北草莓

F. mandschurica Staudt

植株较高，10～25 cm。羽状三小叶，叶正面及背面绒毛均多。叶柄和匍匐茎上具开展绒毛。匍匐茎合轴分枝。聚伞花序，每花序花朵数较多。花序上从第一个分歧处往上绒毛均为紧贴，往下则直立或斜生。花瓣稍叠生或离生，雄蕊花丝较长，高于雌蕊。果红色，圆锥形。果实香味浓，汁液较多。种子黄绿色，凸于果面。宿萼平展或微反折。有时在秋季能再次开花。二倍体2n=2x=14。

The plant is 10~25 cm high. The leaf is trifoliate. There are many hairs adaxially and abaxially on the leaf. The petiole and runner are covered with spreading hairs. The runner is sympodial. The inflorescence is cymose with relatively more flowers. The peduncle is covered with appressed hairs except from the first fork below. The petal is lightly overlapping or apopetalous. The filament is unequal and taller than pistil. The fruit is red, conical, highly aromatic, and juicy. The achene is yellowish green and raised. The calyx is spreading or lightly reflexed. Some accessions of *F. mandschurica* blossom again in autumn. Diploid (2n=2x=14).

东北草莓的植株、叶片、花和果实
Plants, leaves, flowers and fruits of *F. mandschurica*

5.8 中国草莓

F. chinensis Lozinsk.

植株高5~15 cm。羽状三小叶或五小叶。叶片较薄，椭圆形或倒卵圆形，近无柄。叶柄上具开展绒毛或无毛。匍匐茎单轴分枝。花序明显高于叶面，具花2~6朵，雄蕊高于雌蕊。归圃栽植时，群体花量大，单株花朵数多。果实浅红色至红色，小，圆球形或圆柱形，略酸或无味。种子浅黄色或棕色，凹于果面。宿萼平展。二倍体2n=2x=14。

The plant is 5~15 cm high. The leaf is trifoliate or quinquefoliolate, elliptic or obovate, nearly sessile. The petiole is covered with spreading hairs, sometimes glabrous. The runner is monopodial. The runner and peduncles are glabrous or covered with appressed or spreading hairs. Often 2~6 flowers per inflorescence. The fruit is pale red to red, small, globose or columniform, no flavor, or lightly acid. The achene is light yellow or brown, extremely sunken. The calyx is spreading. Diploid (2n=2x=14).

中国草莓的植株、叶片、花和果实
Plants, leaves, flowers and fruits of *F. chinensis*

5.9 台湾草莓

F. hayatai Staudt

　　植株健壮，株高5~18 cm。羽状三小叶，叶厚，深绿色，倒卵圆形或近圆形，前端平楔，中心小叶具短柄。叶柄粗壮、红色。匍匐茎粗壮，红色，合轴分枝。花瓣卵圆形或倒卵圆形，明显离生，花瓣基部短爪呈红色。果实红色，圆球形。种子黄绿色，小，凹陷。宿萼紧贴。二倍体2*n*=2*x*=14。

　　该种仅分布于中国台湾。其植株、叶片等形态特征与黄毛草莓相近，以前曾将其定为黄毛草莓的一个变种（*F. nilgerrensis* var. *hayatai*）（Staudt, 1999），但后来观察到，该种植株（包括果实）含有花色素苷、花粉形态特征明显不同于黄毛草莓，因此被定为一个独立种（Staudt，2009）。

　　The plant is robust with a 5~18 cm height. The leaf is trifoliate, thick, dark green, obovate or suborbicular, and cuneate at the apex. The central leaflet has a short petiolate. The petiole is thick and red. The runner is also thick and red, sympodial branching. The petal is oval or obovate, obviously apopetalous, and is red at the base. The fruit is red and round. The achene is yellow-green, small, and sunken. The calyx is clasping. Diploid (2*n*=2*x*=14).

　　It is only found in Taiwan, China. Previously, it was identified as a variety of *F. nilgerrensis* (*F. nilgerrensis* var. *hayatai)* because its plant, leaf and other morphological characteristics are similar to *F. nilgerrensis* except for a few traits (Staudt, 1999), but later it was observed that the plant contained anthocyanin (including the fruit, flower and runner), and pollen morphological characteristics were significantly different from *F. nilgerrensis*, so it is classified as a new separate species (Staudt, 2009).

台湾草莓的植株、叶片、花和果实（陈右人教授提供照片）
Plants, leaves, flowers and fruits of *F. hayatai* (Provided by Prof. Youren Chen)

5.10 峨眉草莓

F. emeiensis Jia J. Lei

植株健壮，株高5~25 cm。羽状五小叶，或三小叶。叶片卵圆形，小叶近无柄或具短柄。叶柄绿色或略浅红色，较粗，叶柄上绒毛紧贴，或稀少近无毛。匍匐茎红色，单轴分枝，匍匐茎上绒毛向前紧贴、稀少。花序明显高于植株，每花序有1~6朵花。花序梗及小花梗绒毛向上紧贴、稀少。花两性，花药大，花丝明显不等长。花瓣白色、长卵形，离生或相接。果实红色，较大，不规则卵圆形、半球形、短圆锥形。汁液少，无香味，略酸。种子浅红色，凹于果面。萼片宽披针形，副萼片披针形，偶2~3裂，果实成熟时萼片平展或微反折。二倍体2n=2x=14。

The plant is robust with a 5~25 cm height. The leaf is pinnately quinquefoliolate, rarely trifoliate, oval, nearly sessile, or with extremely short petiolule on the central leaflet. The petiole is green or slightly light red, thick, glabrous or covered with extremely sparse appressed hairs. The runner is red, monopodial, and covered with extremely sparse appressed hairs. The inflorescence is obviously taller than leaf with 1~6 flowers per inflorescence. The peduncle and pedicel are covered with sparsely appressed hairs. The anther is relatively big, and the filament is unequal in length and taller than pistil. The petal is apopetalous or slightly overlapping. The fruit is red, relatively big, irregular oval, hemispherical, or short conical, little juice, acid or tasteless without fragrance. The achene is red and sunken. The calyx is wide lanceolate and the calycle is lanceolate sometimes with 2 or 3 lobed apexes. The persistent calyx is spreading or slightly reflexed. Diploid (2n=2x=14).

峨眉草莓的植株、叶片、花和果实
Plants, leaves, flowers and fruits of *F. emeiensis*

5.11 东方草莓

F. orientalis Lozinsk.

植株高10~20 cm。羽状三小叶，小叶倒卵圆形、菱状卵形，近无柄，叶背面常紫红色。叶柄上具开展密绒毛。匍匐茎红色，合轴分枝，具开展密绒毛。花序分歧处常具1枚羽状三小叶或2个苞片。每花序上花朵数较多，归圃栽植时常6~13朵，群体花量大。花梗上具向下倾斜或开展绒毛。花大，直径可达2.5~3.0 cm，花瓣圆形，基部具短爪。雄株花药发育正常，雌株花药瘪小退化、花粉极少或无花粉。果短圆锥形、卵球形，红色，有香味。种子平或略凸于果面，宿萼开展。雌雄异株，偶见亚雌雄异株。四倍体2n=4x=28。

The plant is 10~20 cm high. The leaf is pinnately trifoliate, oval. The petiole is green or slightly light red, thick, glabrous or covered with extremely sparsely appressed hairs. The leaf is trifoliate, obovate or rhombic ovate, nearly sessile, covered with thick spreading hairs adaxially, sometimes purplish red abaxially. The petiole and runner are covered with thickly spreading hairs, and the peduncle is covered with inclining downwards or spreading hairs. The runner is sympodial. Often there is 1 pinnately trifoliate leaf or 2 bracts on the fork of inflorescence. The inflorescence has more flowers, often 6 to 13 flowers per inflorescence in ex-situ conservation. The flower is large, up to 2.5~3.0 cm in diameter. The petal is round with a short claw at the base. The anthers of the male plant are normal, and the anthers of the female plant are shriveled and small, with little or no pollen. The fruit is short conical, or oval-globose, red, aromatic. The achene is flat or slightly raised on fruit surface. The calyx is spreading. Dioecious, or rarely sub-dioecious. Tetraploid (2n=4x=28).

雌花（左）和雄花（右）
Female (left) and male (right) flowers

东方草莓的植株、叶片、花和果实
Plants, leaves, flowers and fruits of *F. orientalis*

5.12 西南草莓

F. moupinensis (Franch) Card.

植株高5～15 cm。常为羽状五小叶，稀三小叶。小叶椭圆形、长椭圆形。小叶无柄或仅中心小叶具短柄。匍匐茎红色，单轴分枝。叶柄、匍匐茎、花梗上具开展绒毛。花序上花朵数少，常2朵（1～3，有时达4朵）。花瓣卵圆形。雌株花药瘪小，雄株花药大，雄蕊花丝不等长。果实橙红色，卵球形或椭圆形。宿萼紧贴。种子黄色，凹于果面。雌雄异株，偶见亚雌雄异株。四倍体2*n*=4*x*=28。

The plant is 5~15 cm high. The leaf is pinnately quinquefoliolate, rarely trifoliate. The leaflet is elliptical or long elliptical, almost sessile, or with a very short petiole on the central leaflet. The runner is monopodial. The petiole, runner and peduncle are covered with thickly spreading hairs. Few flowers per inflorescence, often 2 (1 to 3, sometimes up to 4). The petal is oval. The anther of female plants is small, the anther of male plants is large, and the filaments of stamens are unequal in length. The fruit is orange-red to light red, oval-globose, globose or elliptic. The achene is yellow and sunken. The calyx is clasping. Dioecious, or rarely sub-dioecious. Tetraploid, or rarely sub-dioecious (2*n*=4*x*=28).

雌花（左）和雄花（右）Female (left) and male (right) flowers

西南草莓的植株、叶片、花和果实
Plants, leaves, flowers and fruits of *F. moupinensis*

5.13 伞房草莓

F. corymbosa Lozinsk.

　　植株高5～20 cm。羽状三小叶或五小叶，倒卵圆形。匍匐茎红色，较细，单轴分枝，具长而稀少绒毛。叶柄、花梗上具开展绒毛。伞房花序，每花序上常2～5朵花。雌株花小，花瓣叠生，花药常瘪小；雄株花药大，高于雌蕊，雌蕊退化。果实红色，卵形，果肉粉白色，汁液少，味淡，略有酸味。种子深凹于果面。宿萼平展。夏季高温时植株易成片枯死。雌雄异株。四倍体2n=4x=28。

　　The plant is 5~20 cm high. The leaf is pinnately trifoliate or quinquefoliolate, obovate. The petiole is monopodial, filiform, covered with long but thin spreading hairs. The petiole, peduncle and pedicel are covered with spreading hairs. The inflorescence is corymbose, often 2 to 5 flowers. The female flower is very small, petal overlapping, anthers small and shriveled; the male flower has large anthers, higher than pistil. The fruit is red, oval, pinkish-white flesh, less juice, light taste or slightly sour. The achene is deeply sunken on the fruit surface. The calyx is spreading. Plants are prone to die for high temperature in summer. Dioecious. Tetraploid (2n=4x=28).

雌花（左）和雄花（右）Female (left) and male (right) flowers

伞房草莓的植株、叶片、花和果实
Plants, leaves, flowers and fruits of *F. corymbosa*

5.14 纤细草莓

F. gracilis Lozinsk.

　　植株纤细、矮小，高3～10 cm。羽状三小叶，叶片很小，倒卵圆形，近无柄。匍匐茎极纤细，单轴分枝。叶柄、匍匐茎、花梗上具稀开展绒毛。花序平于或高于叶面，常1～2朵花，花小，花瓣圆形。果红色，很小，圆球形或卵球形，橙红色，无味。种子红色，极小，极凹陷。宿萼反折或开展。雌雄异株。在野生草莓种类中开花很早。夏季高温时植株易成片枯死。四倍体2n=4x=28。

　　The plant is extremely slender and dwarf, with a 3~10 cm height. The leaf is trifoliate, very small, obovate, nearly sessile. The runner is filiform and monopodial. The petiole, runner and peduncle are covered with sparsely spreading hairs. Very few flowers per inflorescence, often 1 or 2. The flower is very small with round petals. The fruit is red, very small, subglobose or ovoid, tasteless. The achene is red, very small, extremely sunken. The calyx is reflexed or spreading. Dioecious. It blooms very early in all the wild strawberry species. Plants are prone to die for high temperature in summer. Tetraploid (2n=4x=28).

雌花（左）和雄花（右）Female (left) and male (right) flowers

纤细草莓的植株、叶片、花和果实
Plants, leaves, flowers and fruits of *F. gracilis*

5.15 高原草莓

F. tibetica Staudt et Dickoré

植株高3～15 cm。羽状五小叶，稀三小叶。叶片深绿色，椭圆形，近无柄。匍匐茎较粗，红色，单轴分枝。叶柄、匍匐茎、花梗上具紧贴或斜向上的绒毛。花序上花数少，常2朵（1～3朵）。花瓣圆形，边缘有时波状。萼片宽披针形，副萼片披针形。果实橙红色至浅红色，卵球形，果肉浅红色，酸，无香味。种子黄红兼有，凹于果面。宿萼平贴。雌雄异株。四倍体2n=4x=28。

The plant is 3~15 cm high. The leaf is pinnately quinquefoliolate, rarely trifoliate, nearly sessile. The petiole, runner and peduncle are covered with appressed or inclining upwards hairs. The runner is thick, red, monopodial. Very few flowers per inflorescence, often 2 (1~3). The petal is round and wavy at the edge. The calyx is wide lanceolate and the calycle is lanceolate often with lobed apex. The fruit is orange-red to light red with light red flesh, oval-globose, or globose, sour, no aroma. The achene is sunken. The calyx is clasping. Dioecious. Tetraploid (2n=4x=28).

雌花（左）和雄花（右）Female (left) and male (right) flowers

高原草莓的植株、叶片、花和果实
Plants, leaves, flowers and fruits of *F. tibetica*

第六章
Chapter 6

中国野生草莓种类的生境
Habitat of wild strawberry resources in China

　　每个野生草莓种类除了有一定的分布范围外，还都有自己独特的生境。在我国，野生草莓一般生长在山区，在平原地区我们极少见到野生草莓的分布。自然界中，分布于中国的野生草莓一般生长在山区的山坡、林缘、草地、灌丛、沟边、路旁等生境下。有的主要生长在林下（如森林草莓），有的主要生长在林缘（如东北草莓），有的生长在夏季凉爽的沟边或山坡（如伞房草莓）；有的生长在海拔很高的高山草地（如裂萼草莓、西藏草莓），有的则生长在海拔较低的林地（如森林草莓、东方草莓）。有些草莓种类在一些原生境中，还经常与别的种类混生在一起，例如我们野外调查发现，二倍体的中国草莓分别与二倍体的五叶草莓、四倍体的纤细草莓混生在一起；二倍体的黄毛草莓常分别与二倍体的五叶草莓、中国草莓、峨眉草莓混生在一起，也常分别与四倍体的高原草莓、西南草莓等混生在一起。黄毛草莓是与其它野生草莓种类一起伴生最多的一个种类。

In addition to a certain distribution range, each wild strawberry species has its own unique habitat. Wild strawberries generally grow in mountainous areas in China, and are rarely seen in plain areas. In nature, wild strawberries generally grow in the hillside, forest edge, grass, brush, ditch, roadside and other habitats. For example, some species mainly grow in the forest (such as *F. vesca*), some mainly grow in the forest margin (such as *F. mandschurica*), and some grow in the ditch or hillside where it is cool in summer (such as *F. corymbosa*). Some grow in the alpine grassland with the highest altitude (such as *F. daltoniana* and *F. nubicola*), but some grow in the woodland with a very low elevation (such as *F. vesca* and *F. orientalis*). According to our investigations, some wild strawberry species are often mixed with other species in some original habitats in the same site. For example, diploid *F. chinensis* often grows with diploid *F. pentaphylla* and tetraploid *F. gracilis* respectively.

Diploid *F. nilgerrensis*, which is the species that most mix with other species, often grows with diploid *F. pentaphylla*, *F. chinensis*, and *F. emeiensis* respectively, and also with tetraploid *F. moupinensis* and *F. tibetica* respectively.

森林草莓（左）和中国草莓（右）的原生境
The habitats in situ of *F. vesca* (left) and *F. chinensis* (right)

6.1 五叶草莓

F. pentaphylla Lozinsk.

五叶草莓生长在山坡、草地。海拔1000～2300 m。

F. pentaphylla grows on the hillside and grassland. Altitude 1000~2300 m.

五叶草莓的生境
The habitat in situ of *F. pentaphylla*

五叶草莓的生境
The habitat in situ of *F. pentaphylla*

五叶草莓原生境的结果状态
The fruit-setting state in situ of *F. pentaphylla*

五叶草莓的白果和红果类型原生境中混生在一起
The white-fruited and red-fruited types of *F. pentaphylla* grow together in situ.

6.2 森林草莓

F. vesca L.

森林草莓生于林下、林缘、路旁、草丛、山坡。海拔230～2200 m。

F. vesca grows in the forest, on the forest edge, roadside, grass, and hillside. Altitude 230~2200 m.

森林草莓的生境
The habitat in situ of *F. vesca*

森林草莓的生境
The habitat in situ of *F. vesca*

森林草莓原生境的结果状态
The fruit-setting state in situ of *F. vesca*

森林草莓原生境下采集的果实
The fruits of *F. vesca* collected from the habitat in situ

6.3 黄毛草莓

F. nilgerrensis Schlechtendal ex J. Gay

黄毛草莓生于山坡、草地、沟边、路旁、林缘，也可以生长在湿润的草甸沼泽地。海拔800～2700 m。黄毛草莓可以分别与五叶草莓、中国草莓、峨眉草莓、西南草莓、高原草莓等混生在一起。

F. nilgerrensis grows on the hillside, meadow, ditch, roadside, and forest edge, and also in wet meadow and marsh. Altitude 800~2700 m. *F. nilgerrensis* often grows in situ together with *F. pentaphylla*, *F. chinensis*, *F. emeiensis*, *F. moupinensis*, and *F. tibetica* respectively.

黄毛草莓的生境

The habitat in situ of *F. nilgerrensis*

黄毛草莓原生境生长在潮湿的苔藓沼泽地

F. nilgerrensis grows on moist mossy swamps in the habitat in situ.

黄毛草莓原生境生长在山坡地
F. nilgerrensis grows on hillside in the habitat in situ.

黄毛草莓原生境生长在坡地石缝中
F. nilgerrensis grows in the crevice of the slope in the habitat in situ.

黄毛草莓原生境下的结果状态
The fruit-setting state in situ of *F. nilgerrensis*

黄毛草莓原生境下采集的果实
The fruits of *F. nilgerrensis* collected from the habitat in situ

黄毛草莓（右）与五叶草莓（左）原生境中混生在一起
F. nilgerrensis (right) and *F. pentaphylla* (left) grow together in situ.

黄毛草莓（中间开花者）与五叶草莓（四周结果者）原生境中混生在一起
F. nilgerrensis (middle in bloom) and *F. pentaphylla* (surrounding in fruit-setting) grow together in situ.

黄毛草莓（左）与高原草莓（右）原生境中混生在一起
F. nilgerrensis (left) and *F. tibetica* (right) grow together in situ.

黄毛草莓与高原草莓原生境中混生在一起
F. nilgerrensis and *F. tibetica* grow together in situ.

黄毛草莓（右）与中国草莓（左）原生境中混生在一起
F. nilgerrensis (right) and *F. chinensis* (left) grow together in situ.

黄毛草莓（左右）与中国草莓（中间红果者）原生境中混生在一起
F. nilgerrensis (left and right) and *F. chinensis* (middle with red fruit) grow together in situ.

黄毛草莓（右）与峨眉草莓（左）原生境中混生在一起
F. nilgerrensis (right) and *F. emeiensis* (left) grow together in situ.

黄毛草莓（右）与峨眉草莓（左）原生境中混生在一起
F. nilgerrensis (right) and *F. emeiensis* (left) grow together in situ.

黄毛草莓（下、白果者）与西南草莓（上、红果者）原生境中混生在一起
F. nilgerrensis (below, white-fruited) and *F. moupinensis* (upper, red-fruited) grow together in situ.

黄毛草莓（开花者）与西南草莓（红果者）原生境中混生在一起
F. nilgerrensis (flowering) and *F. moupinensis* (fruiting) grow together in situ.

6.4 西藏草莓

F. nubicola Lindl.

西藏草莓生于河谷、草丛、山坡、灌丛、沟边。海拔2500～3900 m。

F. nubicola grows in valley, grass, hillside, scrub, and ditch. Altitude 2500~3900 m.

西藏草莓的原生境
The habitat in situ of *F. nubicola*

西藏草莓原生境的结果状态
The fruit-setting state in situ of *F. nubicola*

西藏草莓原生境下采集的果实
The fruits of *F. nubicola* collected from the habitat in situ

6.5 裂萼草莓
F. daltoniana Gay

　　裂萼草莓生于山坡、灌丛、草地、路旁。它是分布海拔最高的种，达3360～5000 m。

　　F. daltoniana grows on hillside, scrub, meadow, and roadside. It is the species with distribution on the highest altitude, up to 3360~5000 m.

裂萼草莓的生境
The habitat in situ of *F. daltoniana*

裂萼草莓原生境的结果状态
The fruit-setting state in situ of *F. daltoniana*

裂萼草莓原生境下采集的果实
The fruits of *F. daltoniana* collected from the habitat in situ

6.6 绿色草莓

F. viridis Duch.

绿色草莓生于山坡、草地、溪旁草丛、矮灌丛。海拔900～2900 m。

F. viridis grows on hillside, grassland, grass beside the stream, and dwarf scrub. Altitude 900~2900 m.

绿色草莓的生境
The habitat in situ of *F. viridis*

绿色草莓的生境
The habitat in situ of *F. viridis*

6.7 东北草莓

F. mandschurica Staudt

东北草莓生于林缘、草丛、山坡、沟边。我们考察收集的标本海拔为240～1050 m。

F. mandschurica grows on the edge of the forest, grass, hillside, and ditch. We collected specimens at the altitude of 240~1050 m.

东北草莓的生境
The habitat in situ of *F. mandschurica*

东北草莓原生境的结果状态
The fruit-setting state in situ of *F. mandschurica*

6.8 中国草莓

F. chinensis Lozinsk.

中国草莓生于山坡、林缘、草地、路旁。海拔1600～2900 m。

F. chinensis grows on the hillside, grassland, forest edge and roadside. Altitude 1600~2900 m.

中国草莓的生境

The habitat in situ of *F. chinensis*

中国草莓的生境
The habitat in situ of *F. chinensis*

中国草莓原生境的结果状态
The fruit-setting state in situ of *F. chinensis*

中国草莓原生境下采集的果实
The fruits of *F. chinensis* collected from the habitat in situ

6.9 峨眉草莓

F. emeiensis Jia J. Lei

峨眉草莓生于林缘、草丛、山坡、沟边、路旁。峨眉草莓原生境下可以与黄毛草莓生长在一起。我们考察收集的标本海拔为1800～3200 m。

F. emeiensis grows on the edge of forest, grass, hillside, ditch and roadside. *F. emeiensis* can grow in situ together with *F. nilgerrensis*. We collected specimens at the altitude of 1800~3200 m.

峨眉草莓的生境
The habitat in situ of *F. emeiensis*

峨眉草莓原生境下开花结果状态
The fruit-setting state in situ of *F. emeiensis*

6.10 东方草莓

F. orientalis Lozinsk.

东方草莓生于山坡、林缘、路旁、灌丛、疏林下、草丛中。海拔280～500 m。

F. orientalis grows on the hillside, forest edge, roadside, scrub, sparse forest, and grass. Altitude 280~500 m.

东方草莓的生境
The habitat in situ of *F. orientalis*

东方草莓原生境的结果状态
The fruit-setting state in situ of *F. orientalis*

东方草莓原生境下采集的果实
The fruits of *F. orientalis* collected from the habitat in situ

6.11 西南草莓

F. moupinensis (Franch) Card.

西南草莓生于山坡、草地、林下。海拔1400～4000 m。西南草莓原生境下可以与黄毛草莓混生在一起。

F. moupinensis grows on the hillside, grassland, and forest. Altitude 1400~4000 m. It can grow in situ with *F. nilgerrensis*.

西南草莓的生境
The habitat in situ of *F. moupinensis*

西南草莓原生境下的结果状态
The fruit-setting state in situ of *F. moupinensis*

6.12 伞房草莓

F. corymbosa Lozinsk.

伞房草莓生于山坡、草丛、沟边。我们考察收集的标本海拔为1500~2460 m。

F. corymbosa grows on the hillside, grass, and ditch. We collected the plant specimens at the altitude of 1500~2460 m.

伞房草莓的生境
The habitat in situ of *F. corymbosa*

伞房草莓原生境下的结果状态
The fruit-setting state in situ of *F. corymbosa*

6.13 纤细草莓

F. gracilis Lozinsk.

纤细草莓生于林下、草丛、沟边。我们考察收集的标本海拔为2600～3500 m。原生境中纤细草莓可以与中国草莓混生在一起。

F. gracilis grows in the woodland, grass, and ditch. We collected specimens at the altitude of 2600~3500 m. It can grow in situ with *F. chinensis*.

纤细草莓的生境
The habitat in situ of *F. gracilis*

纤细草莓在原生境下的雌株（左）和雄株（右）
Female (left) and male (right) plants of *F. gracilis* in situ

纤细草莓在原生境下开花结果状态
The fruit-setting state in situ of *F. gracilis*

纤细草莓（左）与中国草莓（右）原生境中混生在一起
F. gracilis (left) and *F. chinensis* (right) grow together in situ.

6.14 高原草莓

F. tibetica Staudt et Dickoré

　　高原草莓生于草地、林缘、山坡、沟边、路旁。我们考察收集的标本海拔为2900 ~ 3200 m。

　　F. tibetica grows in grassland, forest edge, hillside, ditch side, and roadside. We collected specimens at the altitude of 2900~3200 m.

高原草莓的生境
The habitat in situ of *F. tibetica*

高原草莓的生境
The habitat in situ of *F. tibetica*

高原草莓的雌株（左）和雄株（右）混生在一起
Female (left) and male (right) plants of *F. tibetica* grow together in situ.

高原草莓原生境下的结果状态
The fruit-setting state in situ of *F. tibetica*

中国野生草莓种类的季相

Seasonal aspect of wild strawberry resources in China

　　不同野生草莓种类在春、夏、秋、冬四季表现不同，在不同的季节中，它们在物候期、植株高矮、表观形态、解除休眠早晚、抗逆性、抗病性等方面，肉眼可以看出不同种类有很大差异，表现出不同的季相特征，这是鉴定某个种类较稳定的特性，可以有效区分比较不同种类，尤其当进行集中迁地保存时更为明显。季相特征还可以进行抗逆性评价。例如，在春季，我们能看到五叶草莓、中国草莓解除休眠很晚，表现为植株一直矮化、最晚解除休眠；在夏季，纤细草莓、伞房草莓不耐夏季高温强光，表现为老株叶片成片枯萎、而只有边缘的葡匐茎幼苗叶片保持绿色，但绿色草莓、五叶草莓、中国草莓则耐夏季高温多湿；在冬季来临时，高原草莓表现为非常抗早霜，而五叶草莓则不抗早霜。东北草莓、东方草莓进入休眠很早，10月上旬即开始矮化休眠。

　　注：①本书中季相照片均在沈阳农业大学野生草莓资源圃中拍摄。露地栽植，10月底冬季来临前采用稻草防寒，早春3月底撤除防寒物。②本书中，春季季相是指撤除防寒物后的早春生长状态（4月中旬），夏季季相指资源盛花期的生长状态（5月中旬），秋季季相指夏季高温多雨后初秋的生长状态（8月下旬），冬季季相指冬季来临时的生长状态（10月底）。

The morphological characteristics of each wild strawberry species is different in spring, summer, autumn and winter. In different seasons, for different species, they show different seasonal characteristics with great differences in phenological period, plant height, appearance traits, dormancy release time, stress resistance, and disease resistance, which can be seen by the naked eye. It is a stable characteristic for a certain species and an effective way to distinguish and compare different species, especially when they were ex-situ preserved together. The seasonal characteristics can be also used to evaluate the stress resistance. For example, in spring, we can see that *F. pentaphylla* and *F. chinensis* release dormancy very late, and their plants dwarfed for a long time. In summer, *F. gracilis* and *F. corymbosa* were not resistant to high temperature and strong sunshine, the leaves of all old plants almost withered and only the leaves of young runner plantlets kept green, but *F. viridis*, *F. pentaphylla* and *F. chinensis* were resistant to high temperature and high humidity in summer. Before winter comes, *F. tibetica* is very resistant to early frost, but *F. pentaphylla* is not. *F. mandschurica* and *F. orientalis* entered dormancy very early, and their plants began dwarfing in early October.

Notes: (1) In this book, the photos of seasonal aspect were taken in the wild strawberry resource repository in Shenyang Agricultural University. They were planted in the open field. Before winter came, at the end of October, straw was used to protect against cold, and was removed at the end of March in early spring every year. (2) In this book, the spring season aspect refers to the state after the removal of winter protection (middle April), the summer season aspect refers to the state in full bloom (middle May), the autumn season aspect refers to the state of early autumn after a hot and rainy summer (late August), and the winter season aspect refers to the state when winter comes (late October).

7.1 五叶草莓

F. pentaphylla Lozinsk.

春季季相
Seasonal aspect in spring

夏季季相
Seasonal aspect in summer

秋季季相
Seasonal aspect in autumn

冬季季相
Seasonal aspect in winter

7.2 森林草莓

F. vesca L.

春季季相
Seasonal aspect in spring

夏季季相
Seasonal aspect in summer

秋季季相
Seasonal aspect in autumn

冬季季相
Seasonal aspect in winter

7.3 西藏草莓

F. nubicola Lindl.

春季季相
Seasonal aspect in spring

夏季季相
Seasonal aspect in summer

秋季季相
Seasonal aspect in autumn

冬季季相
Seasonal aspect in winter

7.4 黄毛草莓

F. nilgerrensis Schlechtendal ex J. Gay

春季季相
Seasonal aspect in spring

夏季季相
Seasonal aspect in summer

秋季季相
Seasonal aspect in autumn

冬季季相
Seasonal aspect in winter

7.5 裂萼草莓
F. daltoniana Gay

5月30日
May, 30

5月30日
May, 30

6月15日
June, 15

7月1日
July, 1

7.6 绿色草莓

F. viridis Duch.

春季季相
Seasonal aspect in spring

夏季季相
Seasonal aspect in summer

秋季季相
Seasonal aspect in autumn

冬季季相
Seasonal aspect in winter

7.7 东北草莓

F. mandschurica Staudt

春季季相
Seasonal aspect in spring

夏季季相
Seasonal aspect in summer

秋季季相
Seasonal aspect in autumn

冬季季相
Seasonal aspect in winter

7.8 中国草莓
F. chinensis Lozinsk.

春季季相
Seasonal aspect in spring

夏季季相
Seasonal aspect in summer

秋季季相
Seasonal aspect in autumn

冬季季相
Seasonal aspect in winter

7.9 峨眉草莓

F. emeiensis Jia J. Lei

春季季相
Seasonal aspect in spring

夏季季相
Seasonal aspect in summer

秋季季相
Seasonal aspect in autumn

冬季季相
Seasonal aspect in winter

7.10 东方草莓

F. orientalis Lozinsk.

春季季相
Seasonal aspect in spring

夏季季相
Seasonal aspect in summer

秋季季相
Seasonal aspect in autumn

冬季季相
Seasonal aspect in winter

7.11 西南草莓

F. moupinensis (Franch) Card.

春季季相
Seasonal aspect in spring

夏季季相
Seasonal aspect in summer

秋季季相
Seasonal aspect in autumn

冬季季相
Seasonal aspect in winter

7.12 伞房草莓

F. corymbosa Lozinsk.

春季季相
Seasonal aspect in spring

夏季季相
Seasonal aspect in summer

秋季季相
Seasonal aspect in autumn

冬季季相
Seasonal aspect in winter

7.13 纤细草莓

F. gracilis Lozinsk.

春季季相
Seasonal aspect in spring

夏季季相
Seasonal aspect in summer

秋季季相
Seasonal aspect in autumn

冬季季相
Seasonal aspect in winter

7.14 高原草莓

F. tibetica Staudt et Dickoré

春季季相
Seasonal aspect in spring

夏季季相
Seasonal aspect in summer

秋季季相
Seasonal aspect in autumn

冬季季相
Seasonal aspect in winter

第八章
Chapter 8

中国野生草莓资源的利用
Utilization of wild strawberry resources in China

8.1 中国野生草莓资源的保护
Protection of wild strawberry resources in China

中国的野生草莓资源分布广泛，群体量较大，在一些地方有较大的群体面积，各地采食或用于售卖的较多。野生草莓资源在一些自然保护区内得到了较好的保护，如贵州梵净山景区、四川峨眉山景区、吉林长白山景区、湖北神农架景区、山西五台山景区、河北白石山景区等。但我们在考察中也见到一些地区在野生草莓原生地，被大量开垦成了农田、林地、茶园，甚至修建成了房舍、游乐设施、道路等，野生草莓资源遭到了一定程度的破坏。今后应加强对我国野生草莓资源的保护，尤其我国一些性状特殊、分布狭窄、群体量少、生境易遭到破坏的野生草莓应得到重点保护，如白果中国草莓、白果裂萼草莓、白果西藏草莓、粉果东方草莓、峨眉草莓等珍稀种类。

Wild strawberry resources in China are widely distributed with a lot of large populations. In some places, the fruits of them are harvested or sold. They have been well protected in some nature reserves or scenic spots, such as Mount Fanjing in Guizhou Province, Mount Emei in Sichuan Province, Mount Changbai in Jilin Province, Shennongjia Forest Region in Hubei Province, Mount Wutai in Shanxi Province, and Mount Baishi in Hebei Province. In some areas, the habitats of wild strawberries have been reclaimed into farmlands, forest lands, tea gardens, and even buildings, amusement facilities, and roads. The wild strawberry resources have been damaged to some extent. In the future, we should strengthen the protection of wild strawberry resources in China, especially for some rare species with special characteristics, narrow distribution, small population, and vulnerable habitat, such as white-fruited *F. chinensis*, white-fruited *F. daltoniana*, white-fruited *F. nubicola*, pink-fruited *F. orientalis*, and *F. emeiensis*.

四川广元很多野生草莓原生地被开垦成了农田
Some wild strawberry habitats in Guangyuan City, Sichuan Province were reclaimed into farmland.

湖北西北部很多野生草莓原生地被改造成了林地（上）或茶园（下）
Many wild strawberry habitats in the northwest area of Hubei Province were converted into forest lands (top) or tea gardens (bottom).

8.2 中国野生草莓采食利用

Utilization for fresh market

　　野生草莓芳香浓郁，酸甜适口，营养丰富，富含维生素、糖类、氨基酸、有机酸、微量元素等人体所必需的营养物质，深受人们喜爱。我国野生草莓分布广泛，全国各地经常见到当地住民、旅游者在山区的路旁、沟边、林缘、坡地采集野生草莓进行食用，例如在四川成都、陕西南郑有很多山民和游客采集野生草莓食用。在一些野生草莓集中分布区，如甘肃岷县、陕西略阳等地的集市上有大量野生草莓销售鲜果。在甘肃岷县市场上售卖的是红果的中国草莓，而在陕西略阳市场上售卖的是白果的五叶草莓。野生草莓在陕西略阳当地俗称"瓢"，而在四川及临近的陕西宁强等地则俗成称"泡"。

Wild strawberry fruits are most favorite for the fragrant, sweet and sour taste, and rich in various nutrients such as vitamin, sugar, amino acid, organic acid, and microelement which are necessary for the human body. Because wild strawberries are widely distributed in China, the residents and tourists are often seen picking fresh wild strawberry fruits to eat on roadside, ditch, forest edge and slope in mountain areas throughout the country, such as Chengdu City in Sichuan Province and Nanzheng County in Shaanxi Province. Plenty of wild strawberry fruits are sold in the markets in some concentrated distribution areas of wild strawberries, such as Minxian

County in Gansu Province and Lueyang County in Shaanxi Province. The species sold in the market in Minxian County, Gansu Province is red-fruited *F. chinensis*, while the species sold in the market of Lueyang County, Shaanxi Province is white-fruited *F. pentaphylla*. The wild strawberry is commonly known as 'Piao' in Lueyang County, Shaanxi Province, while it is commonly known as 'Pao' in Sichuan Province and adjacent Ningqiang County, Shaanxi Province.

四川成都山区当地住民采集的野生草莓鲜果
Local residents pick wild strawberry fruits in Chengdu City, Sichuan Province.

陕西南郑公路边游客采集野生草莓鲜果
Tourists pick wild strawberry fruits at the roadside in Nanzheng County, Shaanxi Province.

甘肃岷县集市上销售的野生草莓鲜果
Wild strawberry fruits were sold at a market in Minxian County, Gansu Province.

陕西略阳集市上销售的野生草莓鲜果
Wild strawberry fruits were sold at a market in Lueyang County, Shaanxi Province.

8.3 中国野生草莓栽培利用

Utilization for cultivation in China

野生草莓在欧洲、美洲有几百年的栽培历史了，例如森林草莓、麝香草莓、弗州草莓、智利草莓等。但我国野生草莓一直没有进行商业化生产栽培，都是直接在野外原生地采食或销售。近几年来，由于人们认识到野生草莓的营养价值、芳香品质及源于大自然的安全果品等优势，我国陕西略阳等地开始了小面积的野生草莓生产栽培，开创了将野生草莓用于生产栽培的先例。野生草莓的生产栽培在我国有巨大的发展潜力，值得大力开发应用。

Wild strawberries have been cultivated in Europe and America for hundreds of years, such as *F. vesca*, *F. moschata*, *F. virginiana*, and *F. chiloensis*. However, in China, wild strawberries have not been used for commercial cultivation. Wild strawberry fruits were picked to eat or for sale only in their habitats in China. Just in recent years, due to the recognition of their nutritional value, aromatic quality and safety derived directly from nature, a small area of commercial cultivation has begun in Lueyang County, Shaanxi Province, which created a precedent for the use of wild strawberries for commercial production. The commercial cultivation of wild strawberries has great development potential in China.

陕西略阳的野生草莓生产栽培的模式主要有两种。一种是在野生草莓原生长地，就地进行简单的除草管理，生长接近野生自然状况，我们称为半野生栽培，这种方式省地、省工、省力、低成本，效果较好。另一种是与露地种植大果栽培品种一样，进行商业化生产，只不过在栽培过程中，由于野生草莓植株较小且葡匐茎抽生能力极强，所以垄上植株数较多，株距任其自然，种植实行地毯式的多年一栽制。

There are two cultivation modes for white-fruited F. *pentaphylla* in Lueyang County, Shaanxi Province. One is carrying out simple weeding management in the original habitat. We call it a semi-wild cultivation mode because it's close to the wild natural condition. This mode can save a lot of labor and cost with a good income. The other is the commercial cultivation just like the modern cultivars grown in the open field, but with a multi-yeared planting system and a carpet style for its smaller plants and the stronger runner propagation ability.

陕西略阳野生草莓的原地半野生栽培
Semi-wild cultivation of wild strawberries in Lueyang County, Shaanxi Province

陕西略阳的野生草莓露地商品化生产栽培
Commercial cultivation of wild strawberries in Lueyang County, Shaanxi Province

8.4 中国野生草莓加工利用

Utilization for processing in China

　　野生草莓不仅可以鲜食，而且可以加工成多种食品。由于野生草莓不耐存放，所以越来越多的果实用于加工，如加工制作成馍、馅饼、罐头、果酒、果酱等。秦巴山区野生草莓分布十分广泛，陕西的略阳、宁强，四川的广元，甘肃省岷县、文县、武都等分布最为集中。陕西略阳利用五叶草莓（白果）加工生产制作的"瓢儿馍"，在当地已经形成了一个重要产业，不仅在当地销售量较大，而且销往全国各地。对于当地住民，他们往往采摘野生草莓用于制作馅饼，将鲜草莓果加入面粉中加水搅拌，可添加少许糖，用油煎制。由五叶草莓制作的馅饼，香甜可口，非常好吃。除一些罐头加工厂外，当地一些住民也采用简易方法生产野生草莓罐头。宁强县草莓酒厂过去曾用野生草莓大量生产草莓酒，但近几十年来，由于野生草莓产量低、人工采摘成本高、售价高，草莓酒几乎都改为利用栽培品种了，但一些当地住民经常将野生草莓放于白酒中制作成野生草莓泡酒供饮用。

Wild strawberry fruits not only can be eaten fresh, but also can be processed into various foods. But they cannot be stored for a long time, so they are usually used for processing as bread, pie, can, wine, jam and so on. Wild strawberries are most widely and concentratedly distributed in Qinba Mountains, including Lueyang County and Ningqiang County in Shaanxi Province, Guangyuan City in Sichuan Province, and Minxian County, Wenxian County and Wudu District in Gansu Province. 'Piao bread', made of white-fruited *F. pentaphylla* in Lueyang County, Shaanxi Province, has formed an important industry in the local area, which is not only sold locally, but also sold throughout the country. Local residents often go to pick wild strawberries to make pies. They add fresh strawberry fruits into flour and stir with water, add a little sugar, and fry with oil. The pie made of white-fruited *F. pentaphylla* is sweet and delicious. In addition to some processing factories, some local residents also use simple methods to produce canned wild strawberries. Ningqiang County Strawberry Wine Factory used wild strawberry fruits to make wine in the past, but in recent decades, due to the low yield of wild strawberries, the high cost of manual picking, and the high price, strawberry wine has almost changed to the use of cultivars. Even so, some local residents often put wild strawberry fruits into Chinese liquor to make 'wild strawberry wine' for drinking.

陕西略阳县城街道上大量销售的瓢儿馍
A lot of Piao bread is sold on the streets in Lueyang County, Shaanxi Province.

四川成都山区住民采集野生草莓加工制作的草莓馅饼
Strawberry pies are made of white-fruited *F. pentaphylla* fruits by local residents in Chengdu, Sichuan Province.

甘肃岷县农民制作简易野生草莓罐头
Farmers in Minxian County, Gansu Province make simple canned wild strawberries.

四川成都山区住民采集野生草莓泡制草莓酒
Residents in Chengdu, Sichuan Province collect wild strawberry fruits to make strawberry wine.

8.5 中国野生草莓远缘杂交利用

Utilization for distant hybridization in China

我国原产野生草莓中，有些资源在芳香、白果、粉果、抗病、抗寒、耐涝等性状方面具有独特的优势，可以加以利用，如黄毛草莓芳香、抗叶病、较抗蚜虫，白果五叶草莓果实芳香、抗叶病，粉果东方草莓抗寒、可溶性固形物可达20%以上。日本学者野口裕司利用中国原产的黄毛草莓与栽培品种杂交，结合染色体加倍技术，培育了具有浓郁桃香味的大果十倍体草莓新品种'桃薰'。国内一些单位如沈阳农业大学、江苏省农业科学院等也开展了野生草莓与栽培品种远缘杂交利用工作，并取得了一些进展。例如沈阳农业大学利用栽培品种与东方草莓杂交得到了十二倍体杂种后代YH15–10（$2n=12x=84$），并利用其与八倍体栽培品种回交培育出了一些十倍体水平（$2n=10x=70$）的大果优质草莓新品系。

Some of the wild strawberry resources native to China have unique advantages in strong aroma, white or pink fruit, disease resistance, cold resistance, waterlogging tolerance and so on. For example, aromatic fruit, leaf disease resistance, aphid resistance of *F. nilgerrensis;* high aromatic fruit and leaf disease resistance of white-fruited *F. pentaphylla;* good cold resistance and high soluble solid content of more than 20% of pink-fruited *F. orientalis.* Japanese researcher Dr. Yuji Noguchi released a new cultivar of large-fruited decaploid strawberry 'Tokun' with a strong peach flavor by distant hybridization of cultivars and *F. nilgerrensis* combined with chromosome doubling technology. Some domestic institutes, such as Shenyang Agricultural University and Jiangsu Academy of Agricultural Sciences in China, have also carried out the distant hybridization of cultivars and wild strawberries native to China. Shenyang Agricultural University obtained a dodecaploid hybrid YH15–10 ($2n=12x=84$) by the cross of a cultivar and *F. orientalis,* and then obtained a lot of decaploid large-fruited and high-quality selections ($2n=10x=70$) by the cross of a cultivar ($2n=8x=56$) and this dodecaploid hybrid.

沈阳农业大学利用野生草莓进行远缘杂交培育的十倍体草莓新品系（$2n=10x=70$）
Some decaploid selections ($2n=10x=70$) obtained from the cross of wild strawberries and cultivars in Shenyang Agricultural University

参考文献
References

[1] 邓明琴, 雷家军. 中国果树志·草莓卷[M]. 北京: 中国林业出版社, 2005.

[2] 雷家军, 代汉萍, 谭昌华, 等. 中国草莓属（*Fragaria*）植物的分类研究[J]. 园艺学报, 2006, 33 (1): 1–5.

[3] 雷家军, 薛莉, 代汉萍. 草莓十二倍体种间杂种的获得及其回交研究[J]. 中国农业科学, 2012, 45 (22): 4651–4659.

[4] 雷家军, 张运涛, 赵密珍. 中国草莓[M]. 沈阳: 辽宁科学技术出版社, 2011.

[5] 张运涛, 雷家军, 赵密珍, 等. 新中国果树科学研究70年——草莓[J]. 果树学报, 2019, 36 (10): 1441–1452.

[6] 赵密珍, 王桂霞, 钱亚明. 草莓种质资源描述规范和数据标准[M]. 北京: 中国农业出版社, 2006.

[7] 中国科学院植物志编辑委员会. 中国植物志（第37卷）[M]. 北京: 科学出版社, 1985, 350–357.

[8] Lei J J, Xue L, Dai H P, et al. The Taxonomy of Chinese *Fragaria* species [J]. Acta Hortic, 2014, 1049 (1): 289–294.

[9] Lei J J, Xue L, Guo R.X, et al. The *Fragaria* species native to China and their geographical distribution [J]. Acta Hortic, 2017, 1156: 37–46.

[10] Li C L , Hiroshi I, Hideaki O. Flora of China, Vol. 9 [M]. Beijing: Science Press, 2003: 335–338.

[11] Qiao Q, Edger P P, Xue L, et al. Evolutionary history and pan-genome dynamics of strawberry (*Fragaria* spp.) [J]. PNAS, 2021, 118 (45): e2105431118.

[12] Staudt G. The species of *Fragaria*, their taxonomy and geographical distribution [J]. Acta Hortic, 1989, 265: 3–33.

[13] Staudt G. Notes on Asiatic *Fragaria* species: *Fragaria nilgerrensis* Schltdl. ex J. Gay [J]. Bot Jahrb. Syst., 1999, 121 (3): 297–310.

[14] Staudt G. Strawberry biogeography, genetics and systematics [J]. Acta Hortic, 2009, 842: 71–83.

[15] Zhao M Z, Wu W M, Lei J J, et al. Geographical distribution and morphological diversity of wild strawberry germplasm resources in China [J]. Acta Hortic, 2009, 842: 593–596.